COMPUTER ALGORITHMS AND FLOWCHARTING

GERALD A. SILVER

*Professor of
Business Administration
Los Angeles City College*

JOAN B. SILVER

*Programmer/Editor
Literary Graphics*

Gregg and Community College Division/McGraw-Hill Book Company

New York　　St. Louis　　Dallas　　San Francisco
Auckland　　Düsseldorf　　Johannesburg　　Kuala Lumpur
London　　Mexico　　Montreal　　New Delhi　　Panama
Paris　　São Paulo　　Singapore　　Sydney　　Tokyo　　Toronto

Business forms and illustrations have been used through the courtesy of the following: page 6 (FORTRAN coding form), IBM Corporation; pages 22-23 (flowcharting symbols), American National Standards Institute; page 27 (flowcharting template), RapiDesign, Inc.; and page 28 (flowcharting template), IBM Corporation.

SPONSORING EDITOR: Alix-Marie Hall
EDITOR: Joan Zseleczky
DESIGNERS: Susan Rich and Linda Schaffner
PRODUCTION SUPERVISOR: Gary Whitcraft

COMPUTER ALGORITHMS AND FLOWCHARTING

Copyright © 1975 by McGraw-Hill, Inc. All Rights Reserved. Printed in the United States of America. No part of this publication may be reproduced, stored in a retrieval system, or transmitted, in any form or by any means, electronic, mechanical, photocopying, recording, or otherwise, without the prior written permission of the publisher.

1 2 3 4 5 6 7 8 9 0   WCWC   7 8 4 3 2 1 0 9 8 7 6 5

Library of Congress Cataloging in Publication Data

Silver, Gerald A.
   Computer algorithms and flowcharting
    1. Electronic digital computers — Programming.
2. Flow charts.   I. Silver, Joan B., joint author.
II. Title.
QA76.6.S56   001.6'42   74-26650
ISBN 0-07-057445-6

PREFACE

In 1960 there were less than 10,000 computers installed throughout the world. By 1970 there were more than 100,000 computers installed. Recent studies forecast that by the end of the decade there will be more than 800,000 computers installed in 60 countries. As computer processing continues at an increasing rate to replace the once manual — and later mechanized — processing of data, the need for programmers who can instruct and monitor computers also increases.

However, instructing and monitoring the computer is only part of the programmer's task. Since the business and technical operations handled by computers are numerous and varied, the programmer must be able to analyze the task to be performed by computer and decide on the most efficient means of outlining the instructions for the computer. Once on the job, the programmer soon learns that the time spent in planning a project at its inception is a valuable investment in the future. The programmer is often given inadequate and ambiguous problem descriptions, and it is important that these ambiguities be resolved before the program is written. This means that there is a very great need for the programmer to develop a systematic approach to problem analysis, which is a crucial prerequisite to effective program planning.

Computer Algorithms and Flowcharting is designed to meet this need. It explains in a simple, coherent, and logical way how to analyze a problem and how to structure its solution for computer programming. It describes the algorithms, or sets of logical procedures, that a computer can implement and shows the student or programmer trainee how to illustrate these steps with flowcharts.

The text moves from the simple to the complex. It begins with an explanation of elementary programming theory and flowcharting rules and then gradually presents the move involved, sophisticated programming techniques that are used to solve problems often found in the business and technical data processing environments.

ORGANIZATION AND CONTENT

The text material of *Computer Algorithms and Flowcharting* covers the six steps in the programming cycle — (1) problem analysis, (2) developing algorithms and flowcharting, (3) coding, (4) keypunching and keyboarding, (5) running and debugging, and (6) documentation. It is organized into seven chapters. Chapter One defines the programmer's responsibilities in each step of the programming cycle, describes a computer system in general terms, and discusses the relationship between the programmer and the computer system. Chapter Two takes the student or programmer trainee through the first step of the programming cycle — problem analysis. Chapter Three introduces the student to the American National Standard flowcharting symbols and to proper flowcharting procedures. Chapters Four and Five develop the student's ability to write sound instructions for solving typical business and technical problems by computer and to prepare appropriate system and program flowcharts.

Chapter Six introduces the student to the standard programming techniques necessary for success in programming. These techniques, which include such key programming elements as branches, loops, counters, and arrays, are presented in steps, called *building blocks,* which progress from the simple to the more complex. Chapter Seven encourages the student to apply what has been learned about computer programming to real-life business situations. This is done by presenting a problem and then by showing how to solve it through the use of a technique or a combination of techniques discussed in Chapter Six. Several paragraphs of narrative called *documentation* cite the particular programming techniques that the program solution incorporates. These techniques, together with the carefully drawn and labeled flowcharts, copies of the input and output record layouts, and a description of the procedures each program performs, not only aid students in their understanding of the development of algorithms, but also will be useful to systems analysts or programmers who wish to have at hand a group of basic reference programs as an aid to future programming.

SPECIAL FEATURES

The following special features make this text and reference book truly relevant to the reader who plans a career in systems analysis or computer programming or who simply needs an understanding of programming practices and procedures for a background in another career.

FLOWCHARTING SYMBOLS • The uniform symbols and usage of the American National Standards Institute are employed throughout this material. This is important because the American National Standards Institute is responsible for documenting standards and suggested changes that result from the frequent review of the symbols and terms used in information processing. An American national standard serves as a guide for manufacturers, business and industry, the government, and the consumer.

BUILDING BLOCKS • Fundamental programming techniques, such as counters, arrays, and loops, are presented in 15 modules, or building blocks, which explain the function, applications, and logic of each technique. One or more simple flowcharts serve to illustrate each building block so that the student can learn first how a certain technique solves a simple problem and then how a combination of these techniques solves a more complex problem.

EXERCISE MATERIAL • Each chapter is developed with the awareness that problem-solving and programming skills must be acquired step by step and reinforced continually. Exercises, appearing at the end of Chapters One through Five, after each building block in Chapter Six, and after each unit in Chapter Seven, help reinforce what the student has learned from the preceding material. The exercises in the early chapters concentrate on the language of systems analysis and computer programming. In the later chapters, after the student has acquired the vocabulary needed on the job, the exercises reinforce key terms and provide practice in developing computer algorithms for particular solutions and in implementing these solutions with conventional programming techniques.

INTENDED USE

Computer Algorithms and Flowcharting can be used by a student having no former background in programming languages. It can also be used concurrently with course work on a particular programming language such as COBOL, FORTRAN, BASIC, or assembler language. Developed in this text are all the computer algorithms and programming techniques that a student or programming trainee must have to solve a business or technical problem and to code the programming instructions in a particular programming language.

INSTRUCTOR'S KEY

An answer key to the exercises in the textbook is available to instructors in education and industry. This key provides answers for all exercises that have one correct answer and offers suggestions for evaluating the programming exercises that have more than one acceptable solution.

Gerald A. Silver
Joan B. Silver

CONTENTS

Chapter One

Introduction to Computer Programming — 1

- THE COMPUTER SYSTEM — 2
- COMPUTER LOGIC — 3
- STEPS IN PROGRAM PLANNING AND DEVELOPMENT — 3
- PROGRAMMING SKILLS — 7

Chapter Two

Problem Analysis — 9

- ANALYZING THE PROBLEM — 9
- DEVELOPING AN ALGORITHM FOR SOLUTION — 12
- PLANNING FLOWCHARTS — 15

Chapter Three

Flowchart Preparation — 21

- BASIC SYMBOLS — 21
- SPECIALIZED INPUT/OUTPUT SYMBOLS — 23
- SPECIALIZED PROCESS SYMBOLS — 25
- ADDITIONAL SYMBOLS — 26
- DRAWING FLOWCHARTS — 27

Chapter Four

Elementary System Flowcharting — 34

- DESIGNING A BUSINESS SYSTEM — 34
- SYSTEM FLOWCHARTS — 37

Chapter Five

Elementary Program Flowcharting — 45

- INTEGRATED PROGRAMS AND MODULAR PROGRAMS — 45
- PROGRAM FLOWCHARTS — 49
- GENERAL PROGRAMMING CONSIDERATIONS — 49
- DESIGNING GRAPHIC OUTPUT — 51
- PREPARING PROGRAM FLOWCHARTS — 52
- SAMPLE PROBLEMS — 53

Chapter Six

Program Building Blocks 59

BUILDING BLOCK 1:	Single-Pass Execution	59
BUILDING BLOCK 2:	Unconditional Branch	60
BUILDING BLOCK 3:	Conditional Branch (Two-Way)	61
BUILDING BLOCK 4:	Conditional Branch (Three-Way)	63
BUILDING BLOCK 5:	Conditional Branch (Multiway)	65
BUILDING BLOCK 6:	Simple Loop	67
BUILDING BLOCK 7:	Counters	69
BUILDING BLOCK 8:	Sequential Loops	72
BUILDING BLOCK 9:	Nested Loops	75
BUILDING BLOCK 10:	Terminating a Loop — The Trailer Record	78
BUILDING BLOCK 11:	Terminating a Loop — The Sum-of-the-Fields Technique	80
BUILDING BLOCK 12:	Terminating a Loop Using a Counter	82
BUILDING BLOCK 13:	Limited Loops — A Language Feature	86
BUILDING BLOCK 14:	One-Dimensional Arrays	91
BUILDING BLOCK 15:	Two-Dimensional Arrays	97

Chapter Seven

Applied Programming Logic

UNIT 1:	Simple Calculation and Report Preparation Problem	102
UNIT 2:	Processing Data Files One Record at a Time	105
UNIT 3:	Processing Data Files by Groups of Records	108
UNIT 4:	Classifying Data With Three-Way Conditional Branches	111
UNIT 5:	Preparing Reports With Literal Text and Variable Data	115
UNIT 6:	Processing a Data File With All Records in Storage at the Same Time: Using Arrays	118
UNIT 7:	Using Decision Tables	122
UNIT 8:	Processing Multiple Input/Output Files	128
UNIT 9:	Programming Mathematical Formulas	133
UNIT 10:	Interactive Programming Techniques	137
UNIT 11:	Preparing Graphs	144
UNIT 12:	Performing Numeric and Alphabetic Sorts	148
UNIT 13:	Performing a Binary Search	154
UNIT 14:	File Maintenance Routine	159

Index 168

Chapter ONE
Introduction to Computer Programming

"Mrs. Sanudo, Smithson Company needs 1,000 more assemblies as soon as possible, so would you please get the order rolling?"

Mrs. Sanudo goes right to work, mentally planning and organizing the steps she will follow: "First I'd better see which parts are needed and which ones will have to be ordered. Then I'll check with the factory supervisor and set up a time schedule so I can let Smithson know when the assemblies will be ready..."

Then she executes her plan. She locates the Smithson file and determines the kinds and number of parts needed to complete the assemblies. The inventory file tells her which units are available in stock and which are on order awaiting delivery. She locates the names and addresses of the vendors and prepares purchase orders for those parts that are in short supply. Then she records all this information in the inventory file to keep its status current.

Next she draws up a tentative time schedule indicating when each kind of part will be needed in the manufacturing process and when it will be available. Finally, she alerts the supervisor and discusses the scheduling of the employees and machines in the factory, and incorporates this information into her timetable. Now she can notify Smithson Company of the expected date of completion.

Mrs. Sanudo has just finished a data processing problem-solving activity. It involved forming a plan, or program of action, making decisions, performing computations, and sorting and preparing forms and documents. The entire sequence was initiated by a one-sentence instruction from her employer.

Computers, too, can solve data processing problems efficiently and accurately, but they do not have a person's ability to deduce specific and detailed steps from generalized and unspecific instructions. In one sense, the computer is a mechanical robot. It may be fast, persevering, and accurate, but it cannot think for itself any more than an adding machine, abacus, or calculator can. It must be told what to do, step by step. It must receive specific instructions, in sequence, telling where the data it needs will be found, what processing steps to perform, and where and how the results are to be output.

The set of instructions that directs a computer to solve a problem is called a *program*. It is developed by a programmer who has an understanding of how the computer solves problems, what the problem to be solved involves, and how this is communicated to the computer.

The modern electronic computer differs from adding machines and calculators in one significant way. It has provisions for "remembering" or storing the program it has been given. It is this capability that gives the computer much of its power. Once the computer has received the program, it will carry out the instructions as directed, without further assistance from the programmer.

THE COMPUTER SYSTEM

Since some knowledge of the way the computer system is organized is essential when learning to program, a brief overview of the computer and its four major subsystems is given below (see Figure 1.1).

1 INPUT: The input system reads data, programs, or instructions into the computer. It consists of card readers, tape readers, and other devices capable of sensing holes in paper tape or cards, or magnetized areas on tape. It converts these into electronic pulses that are input to the computer.

2 CENTRAL PROCESSING UNIT (CPU): The CPU is the "brain" of the computer and has three sections. The *control unit* directs the entire system, handles the scheduling and timing, and switches devices in and out at the proper time. The *arithmetic and logic unit* performs mathematical operations and makes comparisons of quantities. The *primary storage unit* holds instructions and the data that is being processed.

Data moves about within the CPU as electronic pulses. These pulses (representing information) travel from the arithmetic and logic portion to primary storage and to input and output devices, as directed by the control unit.

3 SECONDARY STORAGE: The secondary storage system of the computer holds instructions, routines, and data that cannot fit into the primary storage section, or that are not needed by the CPU at that time. It consists of tape drives, magnetic disk drives, and other storage media capable of holding millions of letters or numbers. This data is "accessed" or called into the CPU by the control unit whenever it is needed for processing.

4 OUTPUT: The output system of the computer generates typewritten reports, punched cards, or a display on a visual screen, or it records data on media suitable for input to other

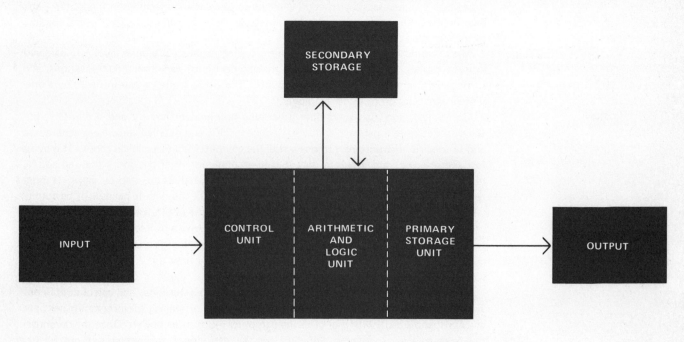

FIGURE 1.1

Computer Algorithms and Flowcharting

machines. It consists of line printers, card punches, video display devices, magnetic tape, or disk drives. These and other devices perform the task of presenting results of computer processing in a usable form or of preparing information suitable for further processing.

COMPUTER LOGIC

A computer can perform only certain types of procedures. It can input (receive) data, output (deliver) data, move data about within the computer system, and make simple logical decisions regarding numeric or alphabetic quantities.

Numeric values can be added, subtracted, multiplied, or divided. These mathematical operations can be combined to perform more complex mathematical processes, such as extracting the square root of a number or finding a tangent or cosine. The computer can also compare numeric values and indicate whether one is less than, equal to, or greater than another.

Alphabetic manipulation is more limited. Groups of alphabetic letters can be compared according to alphabetic order, read in, written out, or moved about within the system.

All of these procedures are combined in different ways by the programmer to solve a data processing problem and to produce usable answers or results.

The capability of the computer to make logical decisions is one of its most important attributes. But unlike humans, its capacity is limited to making only restricted comparisons of data expressed in quantitative terms. A computer cannot distinguish between a "good" and a "poor" employee as a supervisor can. It cannot separate those who are "poor credit risks" from those who are "selected accounts," as a credit manager can. Nor can it sort sales representatives into "below par" people and "top performers."

These qualities must be converted to numeric quantities for computer processing. A computer can sort out employees with production records of over 100 units per hour, send notices to accounts who are delinquent more than 30 days, or prepare bonus checks for sales representatives with sales over $10,000 per month.

Much of the effort of the programmer is directed to reducing qualitative factors to numbers for computer processing.

Using only these rather limited options, the programmer can construct complex programs that solve sophisticated problems. Sometimes it seems more demanding to plan a computer solution to a problem than to carry it out manually. Yet, for repetitive problems, which are common in business, the accuracy and speed of the computer are invaluable.

STEPS IN PROGRAM PLANNING AND DEVELOPMENT

Programmers follow a definite procedure when converting a data processing problem into a computer program. This procedure involves six steps.

1 PROBLEM ANALYSIS: The first step, problem analysis, involves a study of the elements of the problem to be solved. It includes a review of the form and kind of data to be processed and a look at the existing limitations and constraints. The programmer must indicate the following: what the desired results or solution should be; what kinds of information are to be processed; how and where the data is to be manipulated; what computations, equations, variables, and tables must be employed to solve the problem; what results are expected; whether forms need to be printed out by the computer; and what information the user will finally receive from the computer and how it will be displayed. In problem analysis all variables and factors in the problem are defined and reduced to quantitative terms that the computer can handle. The programmer also ascertains what

FLOWCHART OF PAINTING A ROOM

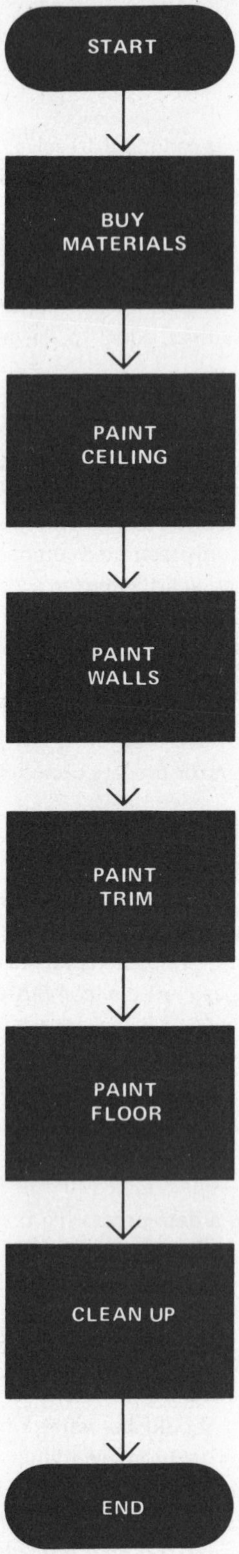

START → BUY MATERIALS → PAINT CEILING → PAINT WALLS → PAINT TRIM → PAINT FLOOR → CLEAN UP → END

FIGURE 1.2

computer equipment will be available to process the data, and what time factors may be involved.

Sometimes, after completing the analysis, the programmer decides that the problem can best be solved by manual means. It may be easier and faster to use a pencil and paper to perform the processing than to write a program for the computer. If a computer solution appears feasible, then the programmer moves to the next step.

2 ALGORITHMS AND FLOWCHARTING: After the programmer has defined the problem in quantitative terms and decided on the output needed, a strategy must be mapped out to reach that end. The logic or sequence of steps the computer must follow to solve the problem is selected. This strategy is called an algorithm. An *algorithm* is a precise set of well-defined rules or procedures for the solution of a problem in a finite number of steps.

People use algorithms every day to solve problems. Mrs. Sanudo developed and followed one when she processed the order for Smithson Company. A homemaker uses an algorithm (recipe) in preparing a fancy dish for the table. A child may follow an algorithm in building a Tinkertoy space shuttle. In business, algorithms are used to balance inventories and reconcile checkbooks and compute income and sales.

An algorithm may be very simple, moving only once through a few procedures, or it may be very complex, involving hundreds of steps, computations, branches, options, and repetitions. The programmer may often experiment with more than one algorithm before selecting the one that will best solve the problem.

The programmer may use several aids in devising and recording an algorithm. One important aid is a flowchart. A *flowchart* is a visual outline of the algorithm in which the steps and processes to be followed are represented by symbols. A flowchart is a graphic representation for the definition, analysis, or solution of a problem in which symbols are used to show operations, data flow, equipment, etc.

Figure 1.2 is a flowchart of the steps a painter might follow in painting a room. Figure 1.3 is a flowchart of a business data processing problem. Details on how to draw flowcharts will be discussed in greater depth in Chapter Three.

The programmer will often use several types of visual devices to illustrate the relationship of the parts. If the problem is very complex, involving many areas in an organization, a *system flowchart*, which shows the relationship of all parts of the organization to the flow of data, is prepared. The system flowchart shows work stations, people, machines, forms, and departments.

The programmer may use a *modular program flowchart* to rough out the major blocks, or procedures, in the algorithm in order to get an overview of the program. Later, each of the major blocks, or modules, is reduced to a series of discrete, executable steps in a *detail program flowchart*.

Another important device the programmer uses to illustrate algorithms is the *decision table*. This is similar to the tables and charts used in magazines and newspapers to show relationships between such things as rainfall figures and geography, production and income, and income and income tax. Decision tables list all contingencies that are to be considered in the description of a problem, together with the actions to be taken. They are of particular use in algorithms with many branches because they illustrate all the possible different paths that can be followed. They will be discussed in more detail in Chapter Two.

3 CODING: After an algorithm has been selected and flowcharted, the programmer must communicate this plan to the computer. The steps in the algorithm must be converted into a list of instructions, or program, in a code or language that the computer can understand. This process is called *coding*.

Computers are composed of electronic devices and components that are either on or off — conducting or nonconducting. Each electronic pathway formed by the conducting elements causes the computer to perform a particular process. The on and off states of the electrical components are represented by a code of ones (1s) and zeroes (0s), called *machine language*. Each different instruction has its own pattern of 1s and 0s.

Computer Algorithms and Flowcharting

Directing the computer in machine language is a time-consuming, tedious, and difficult task. To facilitate communication with the machine, other programming languages, easier for the programmer to use, have been developed. They are called *problem-oriented languages (POLs)*. Special programs, called *compilers*, automatically translate the instructions from the POLs into machine-language instructions without any direction from the programmer. There are many POLs in use today, including FORTRAN, BASIC, COBOL, APL, and PL/I.

The form of the programming instructions varies from one language to another, since each language has its own rules of spelling, syntax, punctuation, and style of commands. Programmers often use a coding sheet (Figure 1.4) to facilitate writing this set of instructions in a language. The coding sheet is marked off in lines and columns, conforming to the rules of each language. Each instruction is coded on a separate line on the sheet.

4 KEYPUNCHING AND KEYBOARDING: The next step is to convert each line on the coding sheet into a form suitable for input to the computer system. The most common means is to keypunch each instruction into a separate punched card for input via a card reader. This device converts the punched data into electrical pulses that are entered into the computer system.

Coded instructions may also be keyboarded on a computer terminal and entered directly into computer storage. Regardless of the method used, the purpose of this step is to prepare the program to be input to the computer system.

5 RUNNING AND DEBUGGING: After the program has been keyboarded or keypunched, it is ready for processing on the computer. The compiler (translator program) for the language used automatically converts the coded instructions into machine language in a process called *compilation*. As each instruction is compiled, it is placed, in sequence, in holding slots in the primary storage section of the CPU. Each slot has its own address, and the instruction stored in it can be called out for use by the CPU by referring to this address.

After all instructions have been compiled and stored, the computer is ready to begin following the instructions in the program. This process is called *execution*.

SAMPLE FLOWCHART

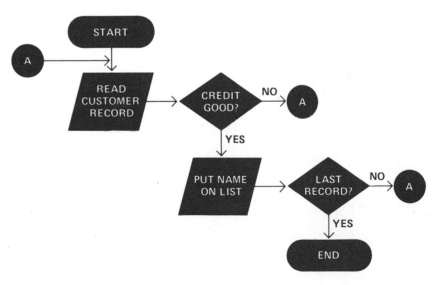

FIGURE 1.3

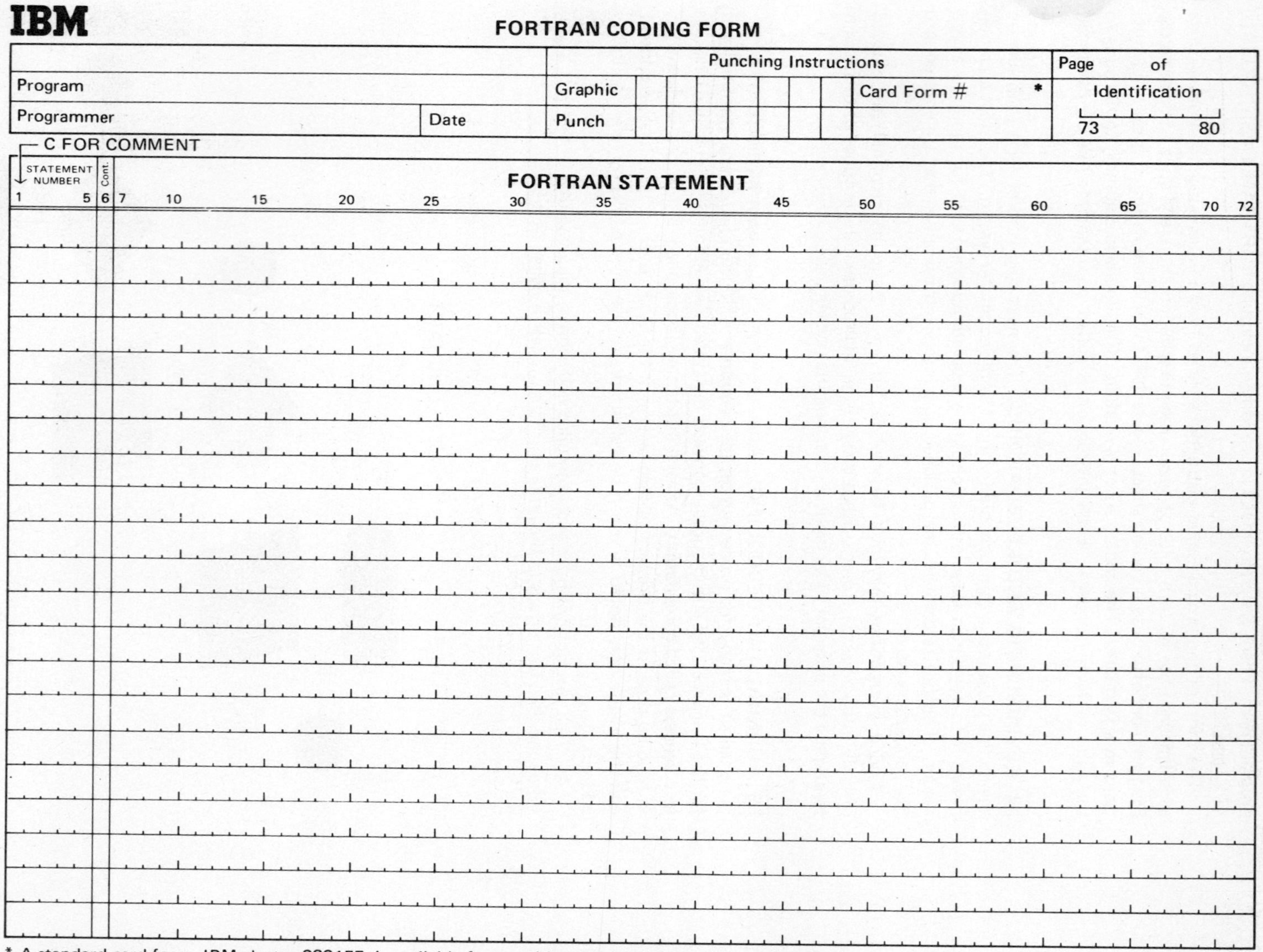
IBM

FORTRAN CODING FORM

Program | Programmer | Date | Punching Instructions | Graphic | Punch | Card Form # * | Page of | Identification 73 80

C FOR COMMENT

STATEMENT NUMBER | Cont. | FORTRAN STATEMENT

1 5 6 7 10 15 20 25 30 35 40 45 50 55 60 65 70 72

* A standard card form, IBM electro 888157, is available for punching source statements from this form.

FIGURE 1.4

The CPU calls the first instruction from storage and performs the action as directed. Then it calls the next instruction and carries it out. The process repeats itself until the program is completed. The computer will always execute instructions in the sequence in which they are stored, unless the programmer directs it to branch or repeat a group of instructions.

At various points in the execution, the programming instructions will direct the computer to locate the data to be manipulated by the program. Data for processing may be made available via an input device, from secondary storage media, or it may be written into the program.

The instructions will also direct the machine to output the results of manipulation and processing via one of the output devices.

After the computer has executed the last instruction in the program, the machine stops or proceeds to the next program in line.

In actual practice, few programs compile and run the first time through the computer. Most contain errors of one type or another. These errors, called *bugs,* may be a failure to follow a language rule, a mistake in the logic followed, or a keyboarding error.

Any errors present must be eliminated, since a program that does not run, or produces erroneous results, is valueless. The operation of removing logic and clerical errors in a program is called *debugging.*

The programmer will also want to run a program with a trial set of data and compare the results to known or manually computed answers. Data designed to cause the computer to execute all branches and decision points in the program will also be used.

6 DOCUMENTATION: After all errors have been eliminated and the program is running correctly, it is ready to be documented. *Documentation* involves preparing a file on the program that contains a list of the program instructions, flowcharts, and a written narrative describing the algorithm and logic followed in the program. The file also contains copies of the input and output record layouts, any decision tables, and a description of all procedures that a program can perform. Documentation completes the preparation of a computer program.

This text emphasizes the processes of problem analysis, algorithm development and selection, and the preparation of flowcharts and decision tables. It does not teach coding, program execution, or preparation of documentation.

While all six steps are essential elements in computer programming, a problem cannot be coded until the programmer has a clear grasp of the problem and how to structure its solution. The material explored in this text will provide the foundation needed for the student to move deeper into computer programming.

A knowledge of a computer language is not a prerequisite for this text. It is assumed that the student already knows one language, or will be learning one concurrently with, or following, a study of this text. Many textbooks and manuals are available to help the student learn to write programs in the various computer languages.

PROGRAMMING SKILLS

The advent of modern computer languages has greatly simplified coding and a programmer no longer has to be a mathematician or an electronics expert to be able to direct a computer to solve a problem. A particular type of skill, however, is still required for successful progress in programming computers.

The programmer must be able to visualize and express a problem in discrete steps. The programmer must have the ability to see abstract relationships, to think in quantitative terms, to organize data, and to plan logical steps and routines.

A programmer is really a director of processing who harnesses the speed and accuracy of the computer to solve problems. The programmer organizes the sequence of steps to be followed and lets the computer perform the complex, tedious, and repetitious computations and routines.

The combination of the human mind and the electronic computer makes an ideal arrangement. People construct the pattern for a solution and the machine carries it out quickly, efficiently, and exactly as instructed.

Exercises

1. List and briefly summarize the four major subsystems of the computer.
2. How is a flowchart an aid to the programmer?
3. List three types of flowcharts often used in programming.
4. Summarize the contents of a documentation file.
5. List several problem-oriented languages.
6. Summarize the function of the keypunching and keyboarding operations.
7. List several skills useful to the programmer.
8. Define the term *program*.
9. Describe some of the kinds of logic the computer can follow.
10. Define an *algorithm*.

Chapter Two
Problem Analysis

The American National Standards Institute (ANSI) defines *problem analysis* as "the methodical investigation of a problem and the separation of the problem into smaller related units for further detailed study." It is an essential aspect of computer programming, yet it is probably the area that is most often the least understood or documented.

Students, as well as professional programmers, sometimes jump immediately into coding before performing problem analysis. The programs that result from this haste often contain clerical and logic errors. Debugging, to eliminate these errors, can cost considerable programming and machine time. And, in the process of debugging, the programmer inevitably has to analyze the problem in order to discover what went wrong.

ANALYZING THE PROBLEM

How does one perform problem analysis? And how does one structure the results of problem analysis for computer solution? This chapter will explain these steps. It will illustrate how to convert a data processing problem that is not clearly expressed into discrete units, suitable for flowcharting and coding.

1 DEFINE THE PROBLEM: The first step in problem analysis is to define the problem. ANSI says problem definition is associated with both the statement and solution phases of a problem and is used to denote the following:

- A. Statement of a problem
- B. A description of the method of solution
- C. The solution itself
- D. Transformation of data
- E. Relationship of procedures, data constraints, and environments

In simple terms, the programmer must determine precisely what transformations, or changes in data, are to take place, how results are to be reported or output, and how the data is to be manipulated in order to achieve these results.

The programmer must be familiar with the basic variables in the problem, including what data is to be processed, where it is, what form it is in, and how much of it is to be used. The programmer must know what the specifications and limiting factors affecting the problem are, as well as what formulas, relationships, rates, and values will be needed to manipulate the data. One should understand how these procedures relate to the logic and arithmetic capabilities of the computer. The programmer must know also what form of output will be needed and what facilities are available.

After the programmer has described the problem, along with its related specifications, factors, and limits, it is time to begin the next step in problem analysis.

2 DEFINE ALL VARIABLES AND REDUCE THEM TO QUANTITATIVE TERMS:

All variables, quantities, and measures must be extracted from the general problem description and expressed as precise, numeric, definite terms. The following are examples of general statements:

Order the "right" amount.
Send a notice to the "slow paying" accounts.
Ship the "rush" items first.

They all contain ambiguous variables or quantities that require subjective judgments to be made before they can be processed. Such statements are acceptable in conversation, but they are unsuitable for handling by the computer.

All statements must be expressed in precise, numeric terms that can be measured, compared, or manipulated arithmetically. All quantities must be reduced to units, such as numbers, feet, meters, grams, dollars, pounds, days, seconds, miles per hour, degrees Centigrade, etc.

Here are examples of variables that are expressed both as general and quantitative terms.

GENERAL VARIABLES (Unsatisfactory for computer)	QUANTITATIVE VARIABLES (Satisfactory for computer)
Past due	30 days late
Good credit	No balance due after 60 days
Fast	100 kilometers per hour
High interest	12 percent
Hot	212°F / 100°C
Tall	6 feet, 4 inches
Discount	10 percent off list price
Bonus	$50

3 REDUCE THE PROBLEM TO SPECIFICS:

After all terms and variables have been reduced to quantitative form, the description of data manipulation, processing, and output must be expressed in precise, specific terms.

A person could probably solve the data processing problems below with only those instructions, but the instructions are too vague and generalized for a computer. They must be restated with all ambiguous references clarified and all elements specified.

A. Prepare travel allowances for the sales staff.
B. Mail discount certificates to our best accounts.
C. Compute the payroll for the last period.
D. Check the inventory.
E. Work out finances on the sale.

For example, in the first statement, the programmer must ask: What is meant by "prepare travel allowances?" Is personal travel to be included or just business-related travel? Are costs other than the cost of operating an automobile used for company business to be included? Are the figures to be output as additional paychecks or recorded and then added to the regular paychecks? Are all salespeople to receive checks or just those using their own cars?

The restatements on page 11 describe the problems more specifically. The question of how the details are to be worked out is still left open, but there is no question regarding what conditions, variables, and actions are included in each problem.

A. Prepare checks for those salespeople in payroll classification #21 who travel more than 25 miles per week on sales-related business. Allow 15¢ per mile traveled. For those who drove more than 300 miles per week on sales-related business, add a repair allowance of $25.
B. Mail a coupon, allowing a 20 percent discount on all merchandise purchased between July 1 and July 31, to the following customers: those with account numbers between 1,000 and 5,000, those who purchased goods during the past 180 days, and those with balances of zero.
C. Prepare paychecks for all hourly employees in department 75, using a base hourly rate of $4.00. Deduct all taxes and withholding amounts according to federal and state tables. Check each employee's record for number of hours worked and for union dues, insurance, and credit union payments to be deducted.
D. Check each item in stock and compare it to the standard minimum level shown on the accompanying inventory table. Prepare a list of items in short supply and indicate the amount of each item to be ordered. Prepare another list of the items that exceed the maximum levels shown on the table and show the quantity in excess.
E. Compute the carrying charges on a sale of $500. The down payment is $100, and the balance will be paid over a period of 12 installments at an annual interest rate of eight percent.

4 ESTABLISH RELATIONSHIPS: The next step is to establish the relationship of all elements in the program. The programmer must now define what actions are to be taken under each possible set of conditions.

Decision tables are visual devices frequently used for this purpose. They are tabular forms that show all relationships between conditions and actions. Each relationship is clearly indicated, eliminating ambiguities and possible errors or omissions.

HOW TO DRAW DECISION TABLES. Decision tables are composed of four sections, or quadrants, as shown in Figure 2.1.

CONDITION STUB:	Lists each condition possible
CONDITION ENTRY:	Lists each combination of conditions that may be encountered
ACTION STUB:	Lists each possible action
ACTION ENTRY:	Lists the combinations of actions for each set of conditions

DECISION TABLE

		Rule 1	Rule 2	Rule 3
IF:	CONDITION STUB		CONDITION ENTRY	
THEN:	ACTION STUB		ACTION ENTRY	

FIGURE 2.1

Chapter Two: Problem Analysis

After the decision table has been set up, the programmer can clearly visualize all possible contingencies or rules. The column formed by a set of conditions and the related actions to be taken is called a *rule*. Rules reflect an *if/then relationship*. IF certain conditions are present, THEN certain actions should take place. The programmer will use these rules as a guide during the next stages of developing the program.

The first step in preparing a decision table is to list the possible conditions that may be encountered in the problem. Next, all possible actions that may result from these conditions are listed on the action stub. Now the programmer indicates all possible combinations of conditions by marking YES or NO, or logical symbols (GT, greater than; EQ, equal; LT, less than) in the appropriate places in a rule. The actions for each rule are usually indicated by placing Xs in the appropriate squares.

ILLUSTRATION OF USAGE. Suppose a programmer wants to prepare a decision table for the travel allowance discussed previously in Problem A. First, all possible conditions are listed:

Salesperson, in class 21
Salesperson, not in class 21
Mileage, 25 or more miles per week
Mileage, less than 25 miles per week
Mileage, between 25 and 300 miles per week
Mileage, more than 300 miles per week
Sales-related travel
Travel not sales related

Then all actions are listed on the action stub:

Prepare check.
Pay mileage allowance.
Pay repair allowance.

Figure 2.2 is the decision table for Problem A. All conditions and actions are included. There are seven rules that are possible under the stated conditions. Figure 2.3 is an example of another decision table.

ADVANTAGES OF DECISION TABLES. Decision tables are an excellent means of clarifying and documenting all of the alternatives that may be found in a problem. All branches and decisions are clearly indicated. Decision tables serve as excellent guides when flowcharting or coding problems. Finally, decision tables are excellent communication tools. Since they clearly illustrate the interrelationships of elements in a program, they facilitate tracing program flow.

DEVELOPING AN ALGORITHM FOR SOLUTION

After the programmer has analyzed and defined a problem, its solution must be planned. A sequence of steps that will input and manipulate the data and produce the desired output is designed. This series of steps is the algorithm.

Usually, there are several algorithms, or strategies, for solving a given problem. The selection of a particular one is based upon such elements as type and availability of data, time and cost factors, and accessibility of equipment.

Algorithms range from the simple to the complex. They may have only a few steps that read in a value, perform a simple procedure, and output the results, or they may be composed of many complicated procedures involving decisions and branches, sorting routines, looping and repetitions, and counting. The programmer may develop one or more decision tables to indicate multiple decisions and branches in an algorithm.

DECISION TABLE FOR PROBLEM A

	1	2	3	4	5	6	7
Classification #21	N	Y	Y	Y	Y	Y	Y
Sales-related travel	–	N	N	N	Y	Y	Y
Mileage, 25 miles or greater	–	N	Y	Y	N	Y	Y
Mileage, 300 miles or greater	–	N	N	Y	N	N	Y
Prepare regular check.		X	X	X	X	X	X
Add mileage allowance.						X	X
Add repair allowance.							X

FIGURE 2.2

DECISION TABLE

	1	2	3	4	5	6	7	8	9	10	11	12	13	14	15	16	17	18	19	20
First bill	Y	Y	N	N	Y	Y	Y	Y	Y	Y	Y	Y	N	N	N	N	N	N	N	N
Extension	N	Y	N	Y	N	Y	N	Y	N	Y	N	Y	N	Y	N	Y	N	Y	N	Y
Business phone	N	N	N	N	Y	Y	Y	Y	Y	Y	Y	Y	Y	Y	Y	Y	Y	Y	Y	Y
Residence phone	Y	Y	Y	Y	N	N	N	N	N	N	N	N	N	N	N	N	N	N	N	N
More than 80 calls	N	N	N	N	N	N	N	N	Y	Y	Y	Y	N	N	N	N	Y	Y	Y	Y
Advertising	N	N	N	N	N	N	Y	Y	N	N	Y	Y	N	N	Y	Y	N	N	Y	Y
Installation charge, $10	X	X			X	X	X	X	X	X	X	X								
Charge per extension, $1		X		X		X		X		X		X		X		X		X		X
Monthly charge, $10					X	X	X	X	X	X	X	X	X	X	X	X	X	X	X	X
Monthly charge, $5	X	X	X	X																
Additional 5¢ per call									X	X	X	X					X	X	X	X
Charge for advertisement							X	X			X	X			X	X			X	X

FIGURE 2.3

Chapter Two: Problem Analysis

An algorithm must express the steps in the solution in a way that will be suitable for computer processing. As discussed in Chapter One, the computer can only perform certain specific operations, and all instructions to it must be composed of one or more of these operations. It can input, output, move data, perform the simple logical operations of comparing quantities and making elementary decisions based upon the comparison of both numeric and alphabetic data, and perform basic mathematical operations.

The symbols which represent the logical operations that the computer can perform are:

- EQ, or, =   Determines whether one quantity is equal to another
- GT, or, >   Determines whether one quantity is greater than another
- LT, or, <   Determines whether one quantity is less than another
- ∪   Determines whether two conditions are present simultaneously
- ∩   Determines whether two conditions are not present simultaneously

The mathematical operations that a computer can perform and the symbols which represent them in computer programming are:

- \+ Addition   * Multiplication   ** Exponentiation
- − Subtraction   / Division

(The specific symbols used to indicate these operations may vary somewhat from one computer language to another.)

These manipulations are combined in various ways to instruct the computer to perform the steps in the algorithm. A computer must be told exactly which of these operations to perform, in what order, and how many times each should be executed. How explicitly these details must be spelled out depends somewhat on the language used for coding.

The programmer takes the procedures and quantitative variables defined in the problem analysis stage and expresses them in terms of these computer-level operations. For example, an algorithm to process the travel allowance for the sales staff (Problem A) might look like this:

INPUT: Each record in the employee file contains the employee's name, classification, miles traveled per week, a code indicating whether travel is sales related, and other information.

OUTPUT: Paycheck on a printed form for each employee eligible for the travel allowance.

ALGORITHM FOR SOLUTION

A. Read in a record from the employee file.
If it is the end-of-file record, terminate execution.
If it is not the end-of-file record, read in name, classification code, miles traveled, and sales-related code.

B. Test classification code.
If it is not equal to 21, go back to read in the next employee record.
If it is equal to 21, continue with the next step.

C. Test miles per week traveled to see if it is greater than 25.
If it is, continue with the next step.
If not, go back to read the next record from the employee file.

D. Multiply the miles per week traveled by 15¢ to compute the basic allowance.

E. Test miles per week to see if it is greater than 300.
If it is not, go to the output routine.
If it is, add $25 to the travel allowance and continue with the next step.

F. Output routine: Print name and travel allowance in appropriate places on the check form.

G. Go back to read the next record from the employee file.

PLANNING FLOWCHARTS

Programmers use graphic outlines, called flowcharts, as they develop their algorithms. In flowcharts, the operations performed and the data flow in the algorithm are represented by symbols. It is far easier to visualize program flow and data manipulation from a flowchart than to interpret paragraphs of written text or programming instructions (see Figure 2.4 on page 16).

THE PURPOSE OF FLOWCHARTS: Flowcharts, very important planning and working tools in programming, have many purposes. They:

- A. Provide communication. Flowcharts are an excellent means of communication. They quickly and clearly impart ideas and descriptions of algorithms to other programmers, students, teachers, computer operators, and users.
- B. Provide an overview. Flowcharts provide a clear overview of the entire problem and its algorithm for solution. They show all major elements and their relationships. They help avoid the possibility of overlooking important details, leaving incomplete branches, etc.
- C. Aid in algorithm development and experimentation. Flowcharts are a quick method of illustrating program flow. It is much easier and faster to try an idea with a flowchart than to write a program and test it on a computer. Experimenting with alternative algorithms is facilitated with flowcharts.
- D. Check program logic. Flowcharts show all major parts of a program. All details in program logic must be clarified and specified. This is a valuable check for maintaining accuracy in logic flow.
- E. Facilitate coding. A programmer can code the programming instructions in a computer language (code) with more ease with a comprehensive flowchart as a guide. The flowchart specifies all the steps to be coded and helps to prevent omissions or errors.
- F. Provide program revision. Flowcharts facilitate modification of running programs; they are an indispensable guide in inserting changes and alterations without disrupting program flow.
- G. Provide program documentation. The flowchart provides a permanent recording of program logic. It documents the steps followed in an algorithm. A comprehensive, carefully drawn flowchart is an indispensable part of documentation for each program.

TYPES OF FLOWCHARTS: The programmer often uses three different types of flowcharts in developing algorithms. They are:

System flowcharts
Modular program flowcharts
Detail program flowcharts

The shapes of the symbols used in these flowcharts have specific meanings. These meanings will be explained in detail in the next chapter.

While the three types of flowcharts are related, each serves a different function.

1. SYSTEM FLOWCHART. The system flowchart gives an overview of the entire business system. It plots the relationship of the data processing problem to be solved to the other parts of the organization. It shows all relevant factors in the network: work stations, forms and documents, elements that input or output data, people, files, communication links, and points of inquiry or access to the system.

Figure 2.5 on page 17 is a system flowchart of an inventory control system used by a large department store. It shows many operations, such as merge and file, and indicates where documents originate and are processed within the system.

DETAIL PROGRAM FLOWCHART
TRAVEL ALLOWANCE — PROBLEM A

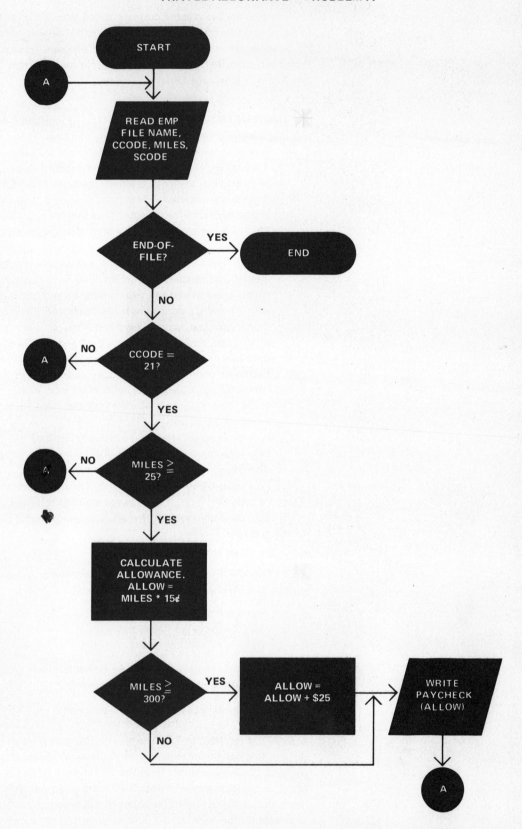

FIGURE 2.4

SYSTEM FLOWCHART

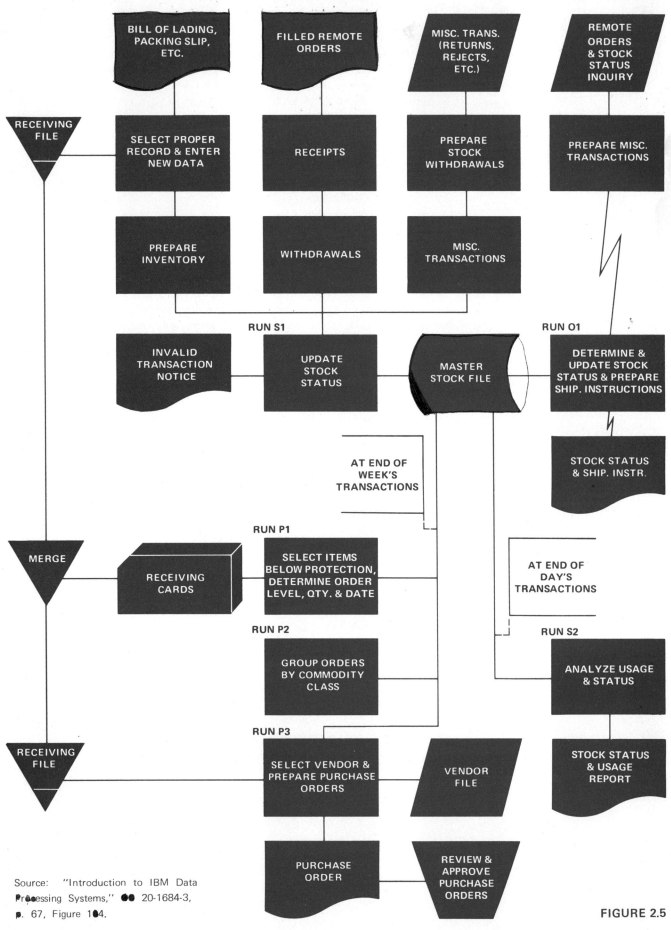

Source: "Introduction to IBM Data Processing Systems," C20-1684-3, p. 67, Figure 104.

FIGURE 2.5

Flowcharts such as this one enable programmers, systems analysts, and managers to see the relationship of all parts of the system. They can see at a glance how and where data originates, and where reports and documents are generated. They get a bird's-eye view of the data flowlines within the company.

It would not be practical for one computer program to perform all of the tasks and functions described in this flowchart. Individual programs can, however, be written to handle the specific data processing operations shown, such as updating stock status, preparing weekly analysis reports, handling inquiries from branches, preparing purchase orders, and spotting slow-moving items.

System flowcharts are often used as a guide when defining a problem and in structuring its solution and output. The programmer will use the flowchart to help determine which functions should be performed by a specific computer program. It will help identify what parts of the system will be involved or affected by changes in data preparation and flow.

2. MODULAR PROGRAM FLOWCHART. Modular program flowcharts are designed primarily to illustrate algorithms for developing and writing a specific computer program. Each block in a modular program flowchart represents a major module or process performed by the program. Specific details on how a given operation is to be done are not included. Only the relationship and sequence of processes are illustrated.

Figure 2.6 is a modular program flowchart of Problem B. It indicates processes such as sort, calculate, and print report. This type of program is advantageous because it allows the programmer to concentrate on designing the major flow of logic in the program and temporarily ignore the computer level details. One may experiment with alternate algorithms without a great expenditure of time or effort. Modular flowcharts are excellent for communicating the major thread of logic followed in a program. The algorithm is not obscured by numerous minute details.

MODULAR PROGRAM FLOWCHART FOR PROBLEM B

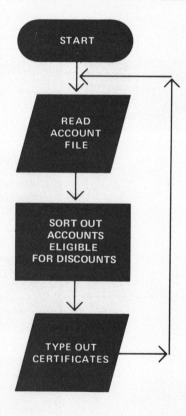

FIGURE 2.6

3. DETAIL PROGRAM FLOWCHART. Detail program flowcharts are the most comprehensive and elemental charts used by programmers. Each module or step in a program is expressed in computer-level terms suitable for coding. Figure 2.4 is a detail program flowchart of Problem A. Figure 2.7 is a detail program flowchart that expands the blocks shown in the modular program flowchart, Figure 2.6, for Problem B.

A detail program flowchart expresses the steps in an algorithm at a level suitable for coding in a computer language. Graphically it shows each executable step. In many instances, each symbol in the flowchart will be represented by one coded instruction in the written program.

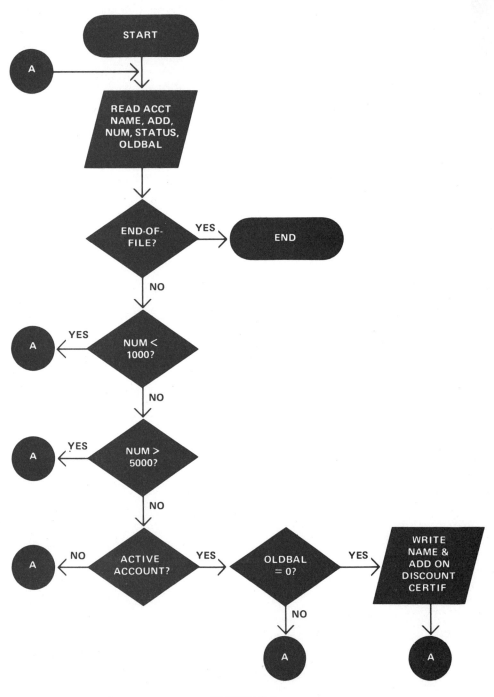

FIGURE 2.7

Chapter Two: Problem Analysis

Exercises

1. Define problem analysis.
2. How do quantities stated as general variables differ from quantitative variables?
3. List three general variables and then restate them as quantitative variables.
4. What are the functions of decision tables?
5. List the four sections on a decision table.
6. Summarize the advantages of using decision tables.
7. Summarize the functions of the system flowchart.
8. Summarize the functions of the modular program flowchart.
9. Summarize the functions of the detail program flowchart.

Chapter THREE
Flowchart Preparation

As we have already seen, flowcharts are composed of symbols of different sizes and shapes. These symbols have specific meanings and functions in a flowchart. The use and design of flowchart symbols have been greatly standardized during the last decade, due mainly to the efforts of the American National Standards Institute (ANSI) and the International Organization for Standardization (ISO).

This chapter explains the common flowcharting symbols shown in Figure 3.1 and their usage. ANSI groups the symbols into three categories: basic symbols, specialized symbols, and additional symbols.

Flowcharts may be drawn using only the basic and additional symbols. The specialized symbols are more precise and specific versions of basic symbols. Their shape conveys the specific media used or operations performed. They give more information since they describe functions in more exact detail than do basic symbols.

The discussion below shows each symbol, gives an example of a typical application, and explains usage.

BASIC SYMBOLS

1 INPUT/OUTPUT SYMBOL: A parallelogram is used to represent an input or output operation. It shows the points in a program where data is made available to a system or where it is output.
Examples: Read part number. Read file on magnetic tape. Punch markup percentage. Print report.

2 PROCESS SYMBOL: A rectangle indicates that a change in the form or value of data will occur, or that an operation or process is to be performed. The process symbol is used to show mathematical calculations, moving data from one location to another, table look-ups, and logic operations.
Examples: Calculate sales tax. Find square root. Compare two values. Copy data from record to another location.

⟶ **3 FLOWLINE:** A straight line between two boxes shows the path of logic flow in a program. It shows the order in which operations are performed and, therefore, indicates the points at which different data has been calculated or processed. An arrowhead on the line shows the direction of flow. Unless indicated otherwise, it is assumed that data flows from top to bottom, and from left to right.

ANSI FLOWCHARTING SYMBOLS

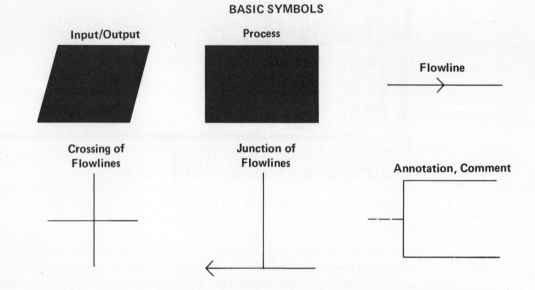

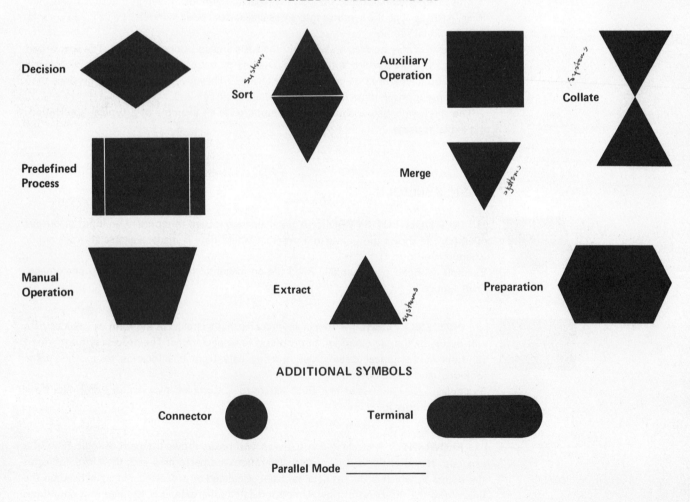

FIGURE 3.1

Flowlines that cross each other are not related. Junctions of flowlines are shown with appropriate arrows indicating the new directions.

4 ANNOTATION, COMMENT SYMBOL: An open-ended box connected to the chart by a broken line is used for descriptive comments, annotations, remarks, or explanatory notes related to the flowchart. The descriptive text is written within the box and the broken line keyed to the appropriate point in the flowchart. The programmer uses this symbol to note formulas, details, or other clarifying comments that will not fit within a symbol in the flowchart.
Examples: Explain a formula used in a calculation. Indicate where data originates. Describe a report.

SPECIALIZED INPUT/OUTPUT SYMBOLS

1 PUNCHED CARD SYMBOL: This symbol, shaped like a punched card, indicates that data is input or output on punched cards. It can represent all forms of data cards, including mark-sense cards, partial cards, stub cards, and mark-scan cards.
Examples: Punch identification numbers into cards. Read cost from cards. Transfer data on mark-sense cards to punched cards.

SPECIALIZED INPUT/OUTPUT SYMBOLS

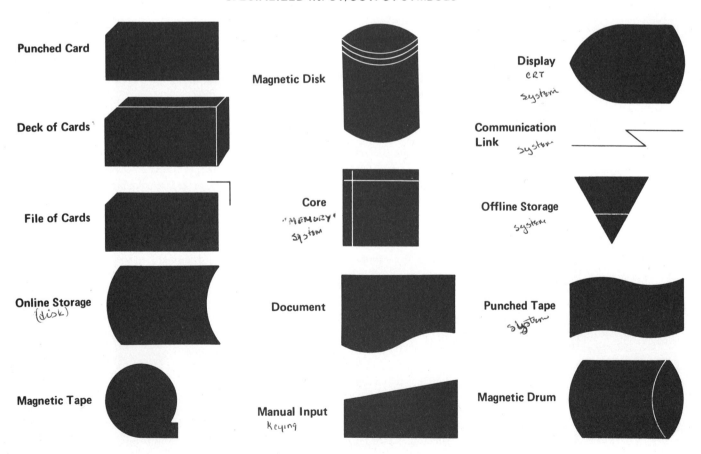

Source: ANSI Flowchart Symbols and Their Usage in Information Processing, X3.5-1970, pp. 15-16.

FIGURE 3.1 (CONTINUED)

A group of cards, such as a program or a data deck, is illustrated by a three-dimensional version of the punched card symbol.

A file of cards containing related data is denoted by a bracket at the upper right of the punched card symbol.

 2 ONLINE STORAGE SYMBOL: This symbol is used to indicate data input or output using online storage media (information stored in devices connected directly to the computer). It represents input/output with magnetic tape, drum, disk, or data cell devices.
Examples: Record discount file on disk. Read selling prices from tape. Call in a subroutine from drum storage.

 3 MAGNETIC TAPE SYMBOL: A symbol shaped like a reel of magnetic tape is a specialized form of the online storage symbol. It shows input or output of data using a magnetic tape device.
Example: Read file of account numbers from tape.

 4 PUNCHED TAPE SYMBOL: A symbol resembling a piece of punched paper tape represents input or output of data using punched tape.
Examples: Punch roster onto tape. Read a file of employee hours worked from tape to computer storage.

 5 MAGNETIC DRUM SYMBOL: A drum or cylinder indicates an input or output function using a magnetic drum storage device.
Examples: A subroutine that calculates markup percentage is stored on a drum. Results of a sorting procedure are recorded onto a drum.

 6 MAGNETIC DISK SYMBOL: This symbol, resembling a stack of magnetic disks, shows that data is to be input or output by magnetic disk.
Examples: Record list of vehicle numbers onto a disk. Read a symbol table from the disk.

 7 CORE SYMBOL: A square box containing a vertical line and a horizontal line shows an input or output operation using magnetic core and other forms of primary storage.
Examples: Load a table into core. Store the results of an update calculation in core. Find and output a file of processed data from primary storage.

 8 DOCUMENT SYMBOL: A symbol that looks like a piece of paper torn from a typewriter is used to show data output in the form of a document.
Examples: Print a credit report. Send a letter to vendors. Print address labels.

 9 MANUAL INPUT SYMBOL: A trapezoid indicates that an input operation will be performed manually while the program is executing. It is used to represent such operations as keyboarding data and setting switches during execution.
Examples: Enter a part number or other piece of data from the computer console. Set console switches when required by the program.

 10 DISPLAY SYMBOL: This symbol indicates that data is to be output on a display device, such as a plotter, a video output device, or a console printer. It may include cathode ray tubes, plotters, or other visual display media.
Examples: Display customer credit standing on a remote video terminal. Display a plot on a plotter device. Print out an error message on the console printer.

11 COMMUNICATION LINK SYMBOL: A symbol resembling a bolt of lightning is used when data is to be transmitted over a communication link. The link may be a microwave circuit, telephone, telegraph, or lease line.
Examples: Transmit a query from a remote terminal to a data base. Transmit data to a remote video display device. Transmit data between two central processing units.

 12 OFFLINE STORAGE SYMBOL: An inverted triangle bisected by a horizontal line is the symbol that indicates that data has been stored offline (not directly connected to CPU). The form of storage can be any media, such as cards or documents, for which access is offline.
Examples: Inventory data should be punched on cards offline. Sales information should be prepared on special OCR (optically read) forms.

SPECIALIZED PROCESS SYMBOLS

 1 DECISION SYMBOL: A diamond shows that a conditional branch has been reached. It indicates that this is a decision point in the program flow and that the program is to select one of several pathways, depending on the condition specified. It may be a two-way branch or have three or more branches.
Examples: Branch to one of two paths depending on whether the temperature is $100^{\circ}C$ or more, or less than $100^{\circ}C$. Branch to path A, B, or C depending on account number.

2 PREDEFINED PROCESS SYMBOL: A process that is described somewhere else is shown by a rectangle with two vertical lines inside. It is used to show that an operation or process that has been previously programmed, such as a subroutine, is to be called in.
Example: Call in a sort procedure to put invoices in sequence by geographic area.

 3 PREPARATION SYMBOL: A hexagon shows that an operation that changes the program, or a sequence of instructions within the program, is to be performed. It may include setting one or more console switches, or initializing a counter or routine.
Examples: Set a counter to zero. Set each value in a table (list of data) to 1.00.

 4 MANUAL OPERATION SYMBOL: A symbol resembling a flowerpot specifies that a manual operation will be performed offline without mechanical equipment. This may include merging two files by hand, manually selecting specific records from a file, or preparing source documents using a pencil.
Examples: Match records in master and detail files. Manually pull records of accounts with overdue balances. Check prepaid items on a list by hand.

 5 AUXILIARY OPERATION SYMBOL: A square indicates that an operation will be carried out offline and not under the direct control of the central processing unit. Such tasks as listing data from punched cards using the line printer and transferring data from cards to tape would be auxiliary operations.
Examples: Punch a duplicate deck of invoice cards. Print out the contents of an employee address file stored on magnetic tape on the line printer.

 6 MERGE SYMBOL: An inverted triangle specifies that two or more files will be merged into a single file.
Examples: Merge wholesale and retail account files into a single file. Merge deposit, withdrawal, and old balance files into one new balance file.

7 EXTRACT SYMBOL: An upright triangle indicates that one or more pieces of data are to be removed from a file. Before extraction, the original file is intact. After extraction, selected items have been pulled from the original file.
Examples: Extract account records with no current balances from the master file. Extract records of items in short supply for reordering.

 8 SORT SYMBOL: Two triangles forming a diamond indicate that data will be put into a new order or sequence. This would include sorting a list of items alphabetically or numerically, or by color, price, etc.
Examples: Sort a sales file into sales per year. Sequence a file of items by part number.

 **9 COLLATE SYMBOL:** The collate symbol is used to indicate that a process of merging with extraction is being performed. Data items from two or more files are rearranged into two or more files.
Examples: Merge records from detail and master files and set aside records without matches. Merge employee records from several departments and extract those eligible for overtime pay.

ADDITIONAL SYMBOLS

 1 CONNECTOR SYMBOL: Two small circles are used to connect separated portions of a flowchart without drawing lines between the parts. They show that a flowline transfers to another point in the chart. These symbols are used when it would be difficult or confusing to draw a flowline between two symbols. One connector indicates where the flow breaks off; the other, where it resumes. The same key letter, word, number, or name is used in both connectors to show their relationship. Arrowheads show the direction. Connectors are also used to indicate the point at which one or more flowlines branch to the same point.
Examples: Connectors may show program flow continuing from the bottom of one column of symbols to the top of the next column. Connectors can indicate the return of program flow to a read routine from several alternate process routines.

 2 TERMINAL SYMBOL: The terminal symbol shows the beginning, end, or interruption points in a program. It always symbolizes the first (START) and last (END) elements on the flowchart. STOP in a terminal symbol indicates the end of a branch and that the program is to terminate execution. INTERRUPT indicates a point in program flow at which the program stops and waits, for example to receive updated input. EXIT indicates that the end of a subroutine or subprogram has been reached and that control should return to the main program.

Some programmers prefer to restrict the use of certain symbols to system flowcharts and the use of other symbols to program flowcharts. Accordingly, system flowcharts are composed of these elements.

    Basic symbols: input/output, process, flowline, annotation
    Specialized input/output symbols: punched card, online storage, magnetic tape, punched tape, magnetic drum, magnetic disk, core, document, manual input, display, communication link, offline storage
    Specialized process symbols: manual, auxiliary, merge, extract, sort, collate
    Additional symbol: connector

Program flowcharts are composed of the following elements.

    Basic symbols: input/output, process, flowline, annotation
    Specialized process symbols: decision, predefined, preparation
    Additional symbols: connector, terminal, parallel mode

DRAWING FLOWCHARTS

The mechanics of combining symbols to produce a finished working flowchart are relatively simple and require only a minimum of skills. A programmer may draw the first flowchart of a program informally, quickly, and sketchily, using a pencil and any available piece of paper. Changes are made by scratching out or erasing. Several quick flowcharts may be drawn in this manner as a program develops. When the programming ideas become crystallized, the programmer usually prepares more carefully drawn, accurate flowcharts suitable for use when coding. A template is usually used for consistency in drawing symbols. When the program is finished and working correctly, an elaborate set of flowcharts is sometimes prepared with India ink and a template.

☐ **MECHANICAL DETAILS:** Using a plastic template will facilitate drawing neat, consistent flowcharts. Several types are available. The template in Figure 3.2 is manufactured by Rapidesign, Inc., and comes in several sizes. All of the symbols conform to the ANSI standard X3.5. The template in Figure 3.3 is manufactured by IBM Corporation (Form No. X20-8020). It contains several additional symbols not found in the ANSI group. The use of either template ensures that symbols will be drawn in a uniform shape, style, and relative size.

RAPIDESIGN, INC., TEMPLATE NO. 54B

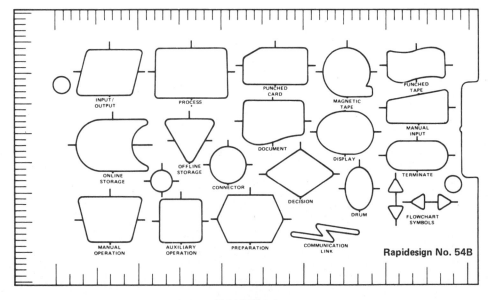

FIGURE 3.2

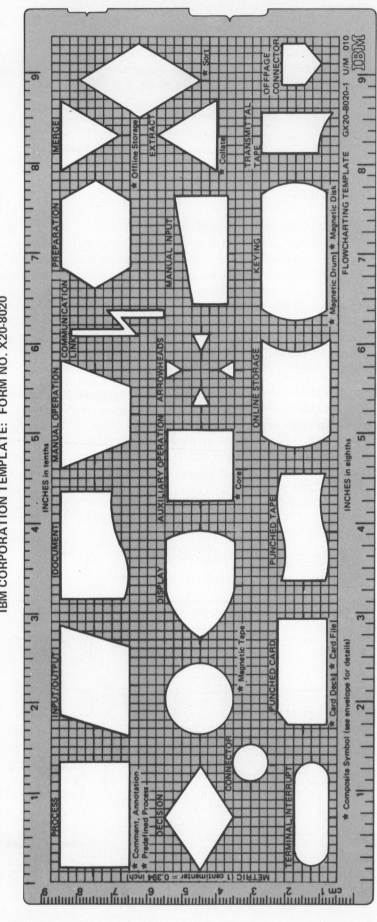

FIGURE 3.3

Source: IBM "Flowcharting Techniques," C20-8152-1, p. 7, Figure 2.

To use a template, the programmer positions it on the page and traces the shape of the symbol needed. If ink is used to draw the flowchart, the template should be raised off the page with small blocks of masking tape to avoid smearing. The programmer may change the size of the symbols, but the dimension ratios should remain as shown in order to conform to the ANSI standards.

Many programmers draw their flowcharts on grid paper. This guides them in drawing parallel lines. Others use flowcharting worksheets produced by IBM (Figure 3.4 on page 30). Plain, unlined paper may also be used. Page size should be uniform, 8½ x 11 or 11 x 17. Each page should be numbered in sequence and labeled with the program name and other pertinent data.

☐ **DESIGNING A FLOWCHART**: Below are some general suggestions that will enable you to prepare neater, more usable flowcharts.

1. PAGE LAYOUT. Use a template when drawing symbols and align the symbols vertically. Leave a minimum of one inch of white space around all pages. Label each page with program name and a brief identification of the routine it performs.

2. SYMBOL SIZE. The ratio of height to width of a symbol is standardized and should not be changed. Templates with symbols of different sizes are available, but the relative dimensions of each symbol must be maintained.

3. DIRECTION OF FLOW. Flow direction is indicated by flowlines that go from one symbol to another. Flow usually moves from top to bottom of the page and from left to right. While not required, it is advisable to terminate each flowline with an open arrowhead to avoid any possible confusion. Arrowheads must be used to show direction on flowlines that move from bottom to top or right to left on a page.

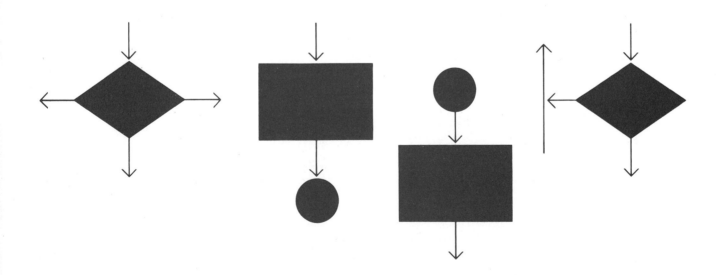

4. SIMPLIFYING A FLOWCHART. Use connector symbols liberally to avoid long or crossing flowlines. This makes it easier to follow program flow and produces neater flowcharts. Always use the simplest and most direct path between two points.

IBM FLOWCHARTING WORKSHEET

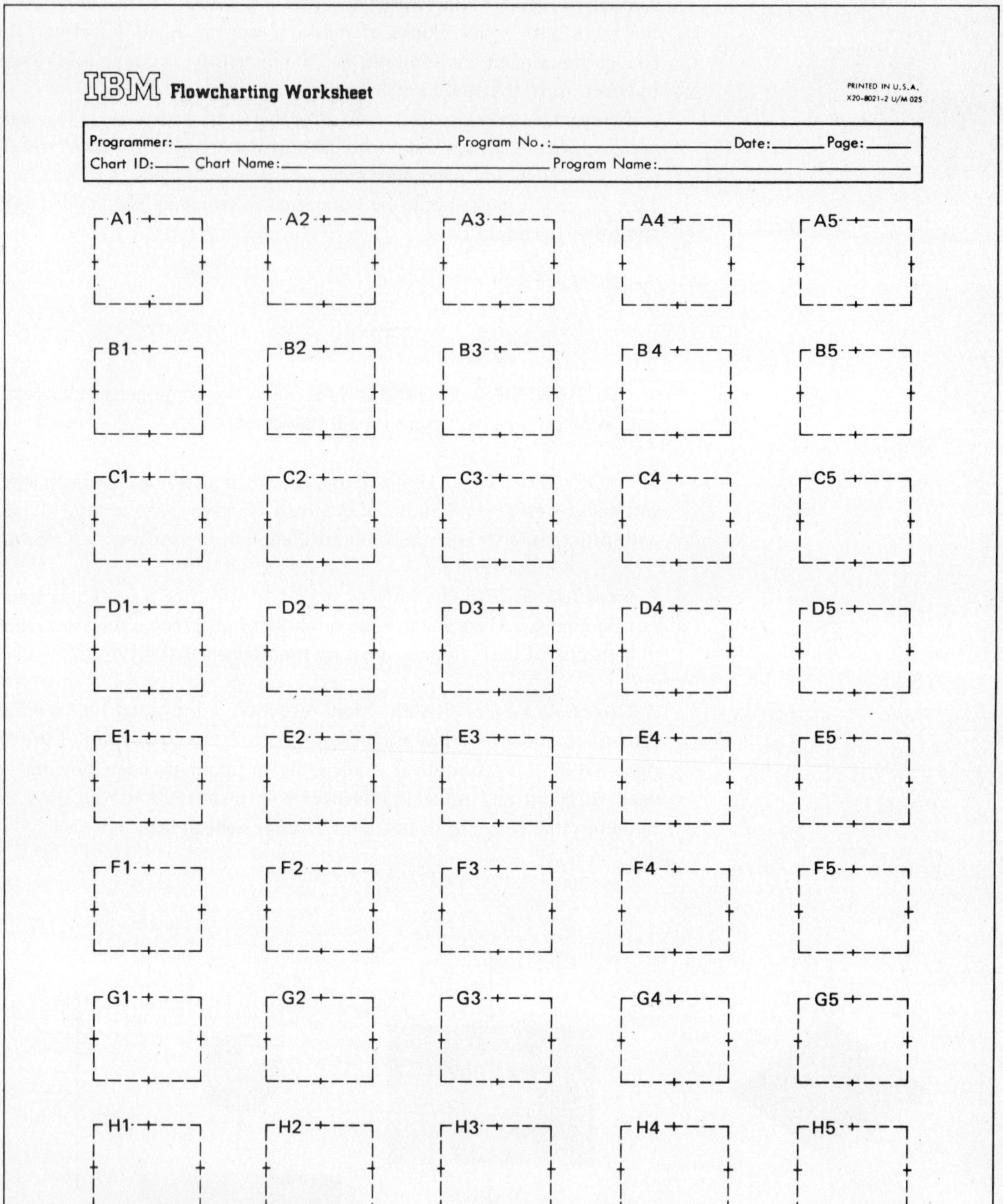

FIGURE 3.4

Source: IBM Data Processing Techniques, "Flowcharting Techniques," C20-8152-1, p. 36, Figure 21

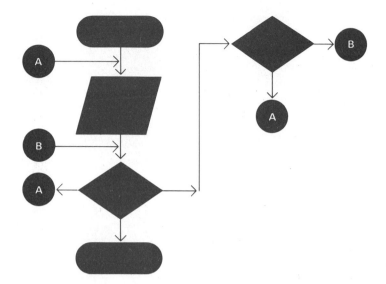

5. JOINING FLOWLINES. Only one flowline should enter each symbol. If flowlines from more than one point must enter a symbol, join them between symbols, using arrows to show direction. More than one line may leave a symbol, and each should be labeled to show when that path would be selected.

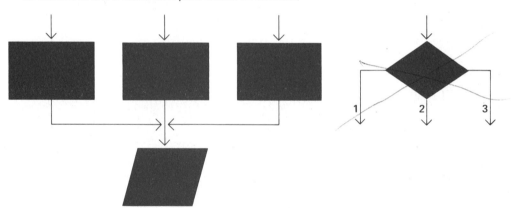

6. PAGE CONTINUATION. Use the connector symbols to show a flowline that extends beyond the page. Use page references above the symbol to indicate the page on which flowline continues or the page from which flowline follows. Label both symbols with the same name to facilitate identification.

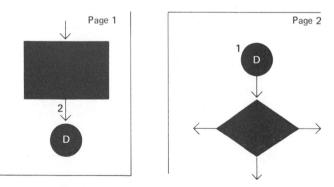

7. LABELING. Label all branches and decision points. It is much easier to follow the direction of flow of a program with labels such as *yes, no, greater than, equal to,* and *less than.* All branches must lead to a specific point in the program that agrees with the program logic. None should be left unresolved. Study the decision points at the top of page 32.

Chapter Three: Flowchart Preparation

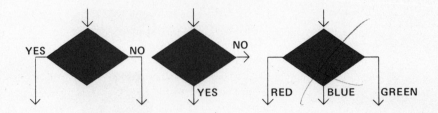

8. MULTIPLE BRANCHES. Decision points with more than three branches may be shown in several ways:

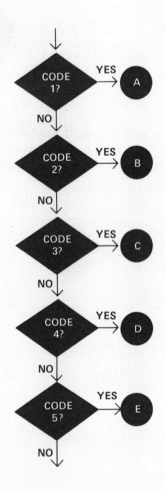

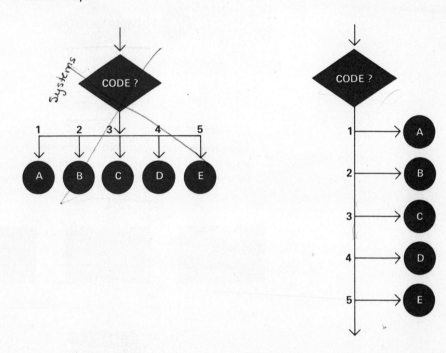

9. DESCRIPTIVE TEXT WITHIN SYMBOLS. Descriptive text should be placed within symbols to explain the steps. Use short, simple, easily understood English words. If more detail is required than will fit within the symbol, use an annotation symbol to elaborate. Avoid using too much text; it can cloud the flowchart with extraneous details.

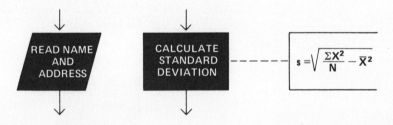

10. HANDLING FORMULAS. If a formula or equation is used in a given step, include it within the process symbol or in an annotation symbol.

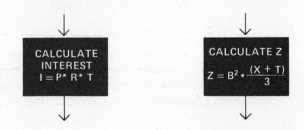

Computer Algorithms and Flowcharting

11. STRIPED SYMBOL. A symbol with a stripe across its top indicates that a subroutine, or modular procedure with many steps, is explained in more detail elsewhere in the flowchart or on another page. A striped process symbol can represent a search, a sort, or a mathematical subroutine. A striped input/output symbol can indicate a procedure that loads a two-dimensional array. The detail flowchart of the subroutine should start and end with a terminal symbol. The striped symbol and the flowchart of the subroutine should be cross-referenced with the same name or number.

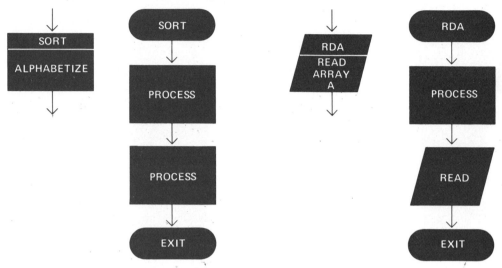

12. REFERENCE SYMBOLS. Cross-reference symbols in the flowchart to statements in the program listing with numbers or reference names. One common method is to place the statement numbers of the coded instructions just above the symbol to the left of its vertical-entry point. This facilitates referencing a step in the flowchart to the coded instruction.

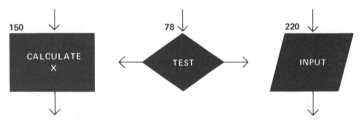

13. CONSISTENCY. Do not mix system flowcharts (which show gross operations and work stations) with program flowcharts (which show executable programming instructions).

Exercises

1. Obtain a flowcharting template. Study it carefully and identify each of the symbols.
2. What mechanical details should be followed when preparing flowcharts?
3. What efforts has ANSI made toward standardizing flowcharting symbols?
4. Draw several flowcharting symbols. Connect them with appropriate flowlines.
5. Draw two columns of flowcharting symbols. Connect them with a connector symbol.
6. Draw a decision symbol with four branches.
7. Use another method to show a decision symbol with four branches.
8. Prepare a striped symbol and the detail flowchart showing the steps in the routine.

Chapter FOUR
Elementary System Flowcharting

A system, as defined by ANSI, is an organized collection of people, machines, and methods required to accomplish a set of specific functions. Business systems are organized to reach certain goals. One of the most important of these goals is to facilitate the flow of data within an organization.

Specialists, called *systems analysts* or *systems engineers,* are given the responsibility of coordinating the components of a data processing system so that it will perform at its maximum level of efficiency. A systems analyst studies and analyzes the ways in which information is generated, processed, and reported within a firm and traces and documents the movement of data through the various elements of a business system.

One of the most valuable tools for this task is the system flowchart. It gives a systems analyst an overview of the organization and helps the analyst to visualize the parts of the business system involved in a particular problem.

This chapter discusses the components and organization of business systems and describes the function of the systems analyst in system design and implementation. The use and applications of system flowcharts in systems analysis are illustrated as well.

DESIGNING A BUSINESS SYSTEM

☐ **DEFINING A DATA PROCESSING PROBLEM:** When designing, studying, or altering a business system, a systems analyst looks at several aspects of the system for the information needed.

1. GOALS. One should define what the goals of a data processing system are and where they fit into the larger goals of the organization.

2. FACTORS THAT CANNOT BE CHANGED. The systems analyst must know which factors in a system cannot be changed. This would include governmental laws and requirements, customer needs, and vendors' levels of support.

3. BASIC COMPONENTS OF THE BUSINESS SYSTEM. One area of concern to the systems analyst is the combination of basic elements that make up a business system—people, methods, and machines. The *people* involved include many types of employees—file clerks, order takers, stock personnel, typists, managers, and salespeople—as well as customers and vendors. The skills these personnel have to offer and their attitudes toward changes in the work procedures strongly affect the success or failure of a new or updated system.

Another major element is the *machines,* or devices, in an organization. Computers, desk calculators, telephones, Teletypes, and typewriters are important factors affecting the efficiency and capabilities of a business system.

The last element in the business system is the variety of *methods* used for processing and moving data. These are the procedures and policies that govern a business's operation. They include methods for inputting and outputting information and generating source documents and reports from computer programs. Uniform policies for handling errors, insufficient or erroneous data, returns, overcharges, and credits must be established.

4. DATA BASE. Another important element of many business systems is a data base. In its simplest form, the *data base* is a collection of data that is stored or maintained by the system. A small retailer, for example, may maintain a list of items in stock. A chain department store, on the other hand, may maintain a more complex data base, with many files of information on personnel, customers, and vendors; financial data; and distribution, shipping, and production details.

In planning the data base, the designer must select both the most efficient method of gathering data and the most suitable storage medium. The type and amount of data to be collected and the file maintenance procedures required to keep the base current must be considered.

It is necessary to study factors involving data access, including personnel needed to access data, frequency with which data is accessed, and method of data access. The systems analyst must decide whether manual card files, tub files, a Cardex system, magnetic disk, drum, data cell storage, or some other system will best do the job in the particular situation.

5. COMMUNICATION LINKS. The systems analyst must design the most effective means of connecting all elements in the system. Work stations must be tied together so that data can move from one place to another quickly, accurately, and conveniently. For some data systems, hand-carried documents, the U.S. Postal Service, or face-to-face communication will suffice. For others, telephone, telegraph, and microwave circuits are necessary. Computerized data processing systems often necessitate the use of telecommunication links.

■ **GOALS OF SYSTEMS ANALYSIS:** The major concern of the systems analyst is to find the optimum arrangement of the components of a business system that can process data promptly, efficiently, and economically. Orderly growth, improvement in the quality, quantity and availability of data, and the reduction of errors are also integral parts of this goal.

In essence, the systems analyst attempts to create a system in which the right kind of data, at the required level of accuracy, is made available to the right place at the right time in the business cycle.

Creating an orderly, efficient business system from these diverse, individual pieces of data is the heart of systems analysis. The analyst must meld these varied facts and opinions into an operating, functioning process that produces the desired goals.

■ **SCIENTIFIC METHOD IN SYSTEMS ANALYSIS:** In systems analysis one must rely heavily upon the principles in the scientific method. The steps in the scientific method are used as a guide for an efficient and objective development of a business system.

1. PROBLEM RECOGNITION. The first step is to establish whether or not there is a data processing problem. Data should:

    A.   Move quickly.
    B.   Be available in the most efficient form.
    C.   Arrive at the right station on time.

The systems analyst may redesign the entire data processing procedure or only a specific area.

When a systems analyst decides there is a data processing problem to be solved, the problem is defined by several criteria. First, the kind of data to be processed—the form it is in, where it originates, and how much is involved—is described. Next, the goals or results are specified, including the form data should be in, who will need it, and where it is to be made available. Then the analyst selects the processing and manipulation procedures to bring the data from its original status to its final format.

2. FACT-FINDING TECHNIQUES. The systems analyst uses several fact-finding techniques to get the information needed to solve a data processing problem.

   A. Interviews and Questionnaires. The systems analyst uses questionnaires and interviews with employees, vendors, customers, and managers to gain information. These people describe their responsibilities and their feelings toward their work and make suggestions for changes or improvements.

   B. Direct Observation. Direct observation is another technique. The systems analyst watches employees work, often analyzing their activities with time and motion studies. Source data and the processes it moves through until it reaches its final stage are examined and recorded. The systems analyst analyzes all documentation, forms, and policies.

   C. Cost and Statistical Analyses. Cost studies and statistical analyses can elicit important information. The systems analyst uses these techniques to investigate and compare costs and efficiency levels in a business system.

3. SYSTEM DESIGN. The next step is to design a system that can meet the requirements specified in the problem-definition stage. The systems analyst thoroughly reviews the existing system and considers the available skills and machines, determining what procedures, operations, and sequences of activities already exist that can be built upon. In deciding upon the most advantageous combination of methods for handling data, the analyst considers the following areas of concern:

   A. Input. Punched card, punched tape, optical character recognition (OCR), and magnetic ink character recognition (MICR) are all possibilities for originating source data.

   B. Output. The systems analyst considers the amount of data to be output and what function it serves. One might want to use video display terminals, line printers, punched tape, magnetic disk, or a combination of these elements.

   C. Processing. Data can be processed by computer, punched-card, mechanical, or manual methods. If computer programs are necessary, the systems analyst decides who should write them, which language to use, and which operations they should perform. The analyst also designs the routines followed in manual and punched-card processing procedures.

   D. Personnel. A systems analyst makes decisions regarding personnel, taking into consideration each person's job and defining tasks and responsibilities. A systems analyst may determine how many typists, clerks, phone operators, programmers, computer operators, and card-punch operators will be needed in a new system. The systems analyst considers whether people with the necessary training to carry through new methods of processing are available, or whether retraining of personnel will be necessary.

Using the results of this analysis, the analyst designs a system that will meet the requirements of the problem. The new plan may involve a complete revision or reordering of the existing data processing system. It may involve only minor changes, such as buying a better typewriter or rearranging a few work stations. It may involve purchasing new equipment and retraining personnel.

The systems analyst experiments with alternative data processing procedures to find the most efficient arrangement of all components. Here system flowcharts are indispensable. They allow the systems analyst to graphically chart data flow and work stations and to conveniently compare alternatives.

4. SYSTEM IMPLEMENTATION. In the next phase, the systems analyst installs the system or modifications that have been designed. Other employees and outside personnel are usually closely involved during this stage. Forms are designed and prepared, equipment is ordered, and new procedures are specified. The firm may hire personnel, install communication lines, or purchase business and data processing equipment.

If new or modified electronic data processing facilities are specified, (and if existing facilities cannot be modernized,) they may have to be purchased. Testing, debugging, and documenting computer programs are often major items in the plan.

5. SYSTEM EVALUATION AND MODIFICATION. The last step in a business system analysis is evaluating the system. In this phase the analyst judges the finished product as compared with the goal set previously.

A. Do the results meet the specific requirements?
B. Are the costs as estimated?
C. Is the level of accuracy adequate?
D. Are the personnel involved satisfied with the new system?

Rarely does a system work perfectly when it is first installed. Usually, modifications and changes must be made. System flowcharts may need to be reviewed or redesigned so that more accurate predictions can be made. The basic components may need to be altered to find a more optimum arrangement of operations. These changes and alterations are designed, installed, and evaluated again on their performance. Only after the system produces the desired results can it be said to be a working system.

The job of the systems analyst does not stop here; systems are not static. Business needs change—new products are introduced and new marketing methods are employed. These changes bring with them new data processing problems to be solved.

SYSTEM FLOWCHARTS

System flowcharts play a vital role in systems analysis. They graphically illustrate the elements and characteristics of a business system and express its structure and relationships in terms of flowcharting symbols.

During the problem recognition stage they illustrate data flow and relationships in the existing system. During the system design step, they allow the analyst to conveniently experiment with alternative data flow plans. They provide graphic records to serve as guides during implementation of a modified or newly designed business system.

During the evaluation stage of systems analysis, system flowcharts reflect the changes and alterations made. They again provide an opportunity for testing alternatives and possible variations. The implemented, working system will be documented in a system flowchart that is carefully drawn, accurate, and comprehensive.

The variety of specialized input/output and processing symbols allows a system flowchart to describe a business system with a high degree of clarity and precision.

☐ EXPRESSING A BUSINESS SYSTEM AS A SYSTEM FLOWCHART: The systems analyst relies on the steps involved in problem analysis when converting a complex business data processing system to symbols on a system flowchart.

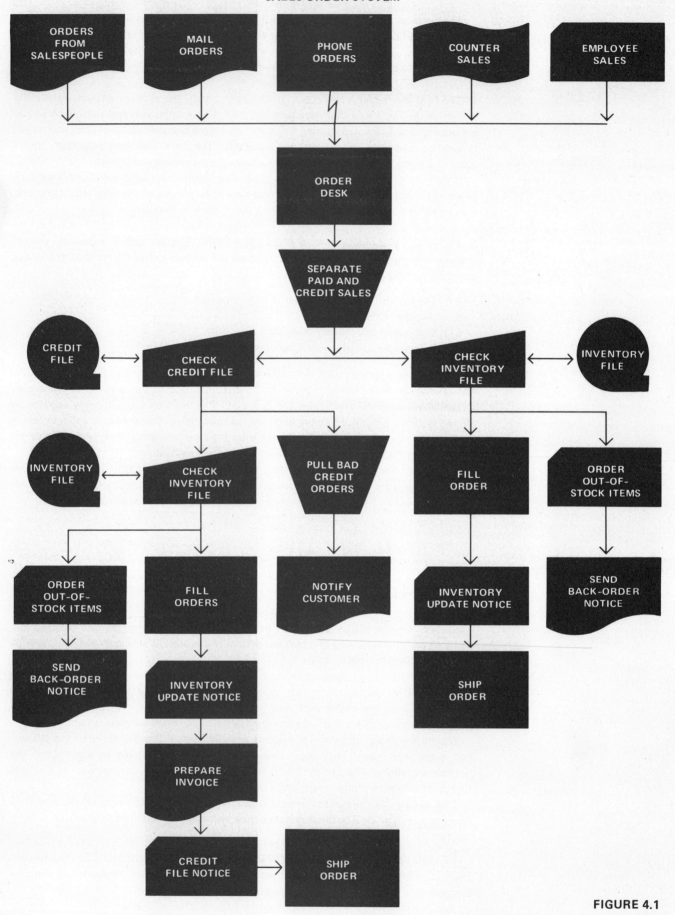

FIGURE 4.1

First the problem or business system must be defined and described. All relevant variables must be specified and expressed in flowcharting symbols. All procedures and processes involved in data manipulation must be shown. The direction and means of movement of data between work stations must be specified and illustrated. Finally, all relationships and alternate pathways must be clearly indicated.

One way to define a business system is to describe it in terms of its fundamental purpose or basic concept: Is it designed to process sales orders, to maintain a data base, to handle debits and credits, or to prepare reports and projections for management?

The elements of a sales order system, for example, are best charted by showing the steps involved in processing an order from its inception to the delivery of the item ordered, or perhaps even to the department where the bill is paid and the accounts receivable file closed. A flowchart for a sales order system would stress such elements as departments (purchasing, sales, inventory) or work stations involved, data transformation, and documents generated.

Another very common data processing system is one organized around a central file, or data base. The flowchart for this system would show the input to the data base, the procedures involved in its maintenance, and the output that is generated. Work stations, format of data input and output, storage and input/output media, and all communication links would be important elements of this flowchart.

☐ EXAMPLES OF SYSTEM FLOWCHARTS:
Following are system flowcharts that illustrate four business systems, each designed around a different basic concept.

1. SALES ORDER SYSTEM. Figure 4.1 illustrates a typical sales order system used by a medium-size firm. Orders are received in various ways, including over the telephone, through outside salespeople, and over the counter.

The orders begin their processing at the order desk, where they are separated into paid sales and credit sales. Paid orders are filled and shipped to the customers and the inventory file is updated. If items are out of stock, this is noted on punched cards and the customer is notified by letter.

Credit sales are first checked for credit rating. Orders from customers with insufficient credit standings are pulled aside; these customers are notified. Orders from customers with good credit are next checked against the inventory file. Items that must be back ordered are noted on punched cards for the inventory department and in letters for the customers' information.

Other orders are filled, the inventory file is updated, invoices are prepared, the credit file is updated, and the order is shipped.

2. CONSUMER CREDIT SYSTEM. Figure 4.2 illustrates a consumer credit system. It is composed of a data base and relies heavily upon remote data inquiry and retrieval. The purpose of the system is to provide up-to-date credit information for a group of participating merchants.

A master credit file (data base) is maintained on a data cell. It is created and maintained by information coming in from many sources. Participating merchants send in reports on customers; court records on bankruptcy proceedings, judgments awarded, and criminal indictments and convictions are monitored for pertinent information.

Details on both sides of a disputed account are also entered into the data base as a protection to the persons listed in the master file. The file is periodically examined so that old records may be deleted.

There are three methods of gaining access to the file. First, customers can call the credit bureau for information. An operator keyboards each request on a remote terminal and transmits it to the computer for processing. The computer searches the credit file and selects up to three possible accounts and displays them on the remote terminal. The operator decides which one is correct and keyboards the selection to the computer. It responds with a display of the proper record. The operator relays this information to the customer by phone. The entire procedure takes two to four minutes.

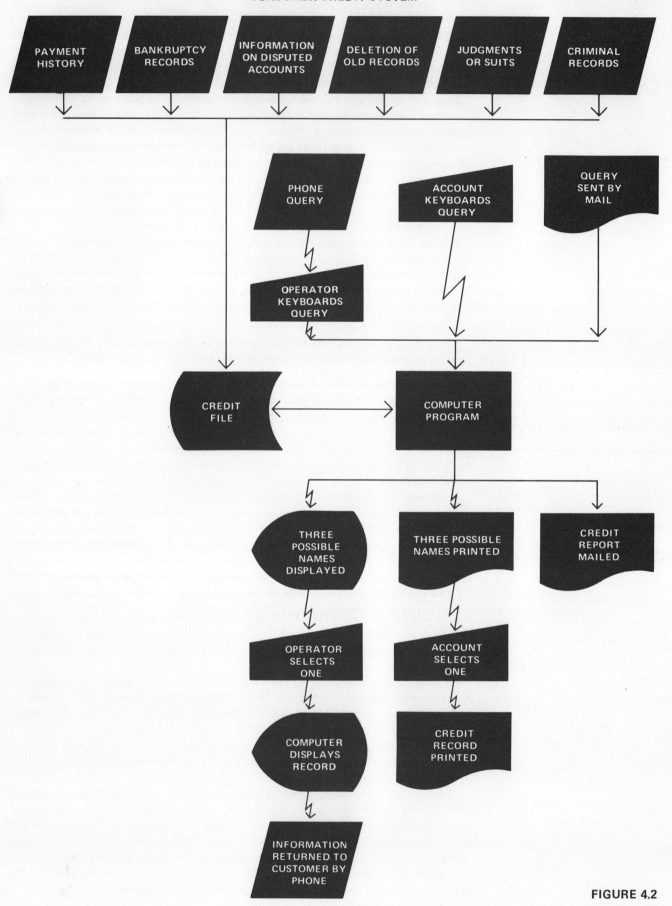

FIGURE 4.2

Computer Algorithms and Flowcharting

A second method of gaining access to the file begins at an account's office with the use of a remote terminal and printer. The account keyboards the request, which is answered by the computer with a printed document.

The third method of gaining access is by mail. Printed documents are mailed back to the customer with the information requested.

3. REAL ESTATE LISTINGS. Figure 4.3 flowcharts a system that produces current, comprehensive real estate listings for member brokers. The heart of the system is a data base. Brokers interview prospective sellers and obtain complete descriptions of their property. This data is punched into cards and input to the computer, where it is stored in the data base. Other input includes data on completed sales and related information on school districts and tax rates.

The computer processes this data and prints out several reports, which are duplicated and sent to all member brokers. A master bimonthly listing shows all property for sale and gives considerable descriptive details on each piece of property. Reports showing new listings are generated and distributed daily.

Weekly, reports on recent sales are prepared and sent to the sales staff to keep their information current. A special report showing prices and descriptions of comparable property by geographic area is made available periodically to help clients establish selling prices for their property.

4. ONLINE FLIGHT RESERVATION SYSTEM. Figure 4.4 illustrates an online flight reservation system that involves a data base, remote data input and output, and multi-processing. The heart of the system is a data base and a computer located at the airline company headquarters. Two smaller onsite computers provide emergency backup services. Remote terminals located in nearby cities and airports access the data base and computer via telephone lines.

Several remote computers located in other countries are connected to the main computer and the data base by communication lines. Each remote computer has its own remote terminals in local airports and cities. (Only one remote system is shown on the system flowchart, but all are similar.)

Reservations and inquiries come to the computer from the telephone and remote terminals. The computers access the data base, update it as new transactions are made, and answer a variety of queries from the terminals. They are programmed to display information on alternate flights and to make notes of special customer needs (such as those of children traveling alone or of handicapped travelers) for flight personnel.

The computers process queries to and from other carriers. The computer at the headquarters is programmed to prepare current documents on flight schedules and passenger lists for the catering departments and movement control. When flights are late, it prepares a list of passengers who will have missed connections and suggests alternate flight schedules.

☐ **PREPARING SYSTEM FLOWCHARTS:** The following rules and guidelines will facilitate drawing system flowcharts and will improve their usefulness.

1. SHOW DATA TRANSFORMATION. Clearly indicate the points at which data originates, the change in form that it undergoes as it moves through a business system, and its final status.

2. SPECIFY WORK STATIONS. Show all work stations and specify the tasks the personnel perform. Show all processing procedures and manipulation of data.

3. USE SPECIALIZED SYMBOLS. The quality of a system flowchart is enhanced if the symbols specify the media and operations as precisely as possible. For example, if a tape record is to be read, use the tape symbol rather than the general input/output symbol.

Chapter Four: Elementary System Flowcharting

REAL ESTATE LISTING SYSTEM

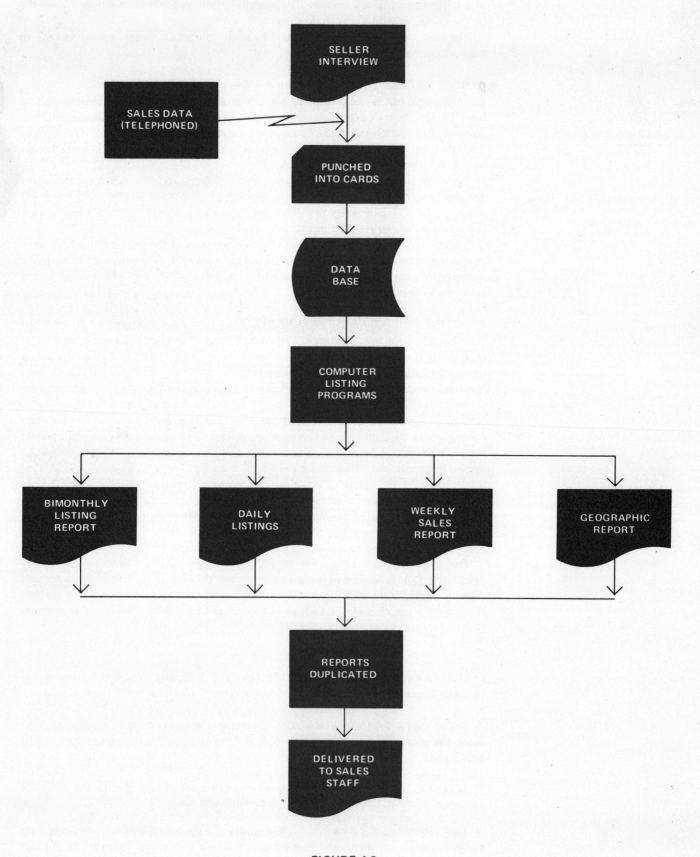

FIGURE 4.3

ONLINE FLIGHT RESERVATION SYSTEM

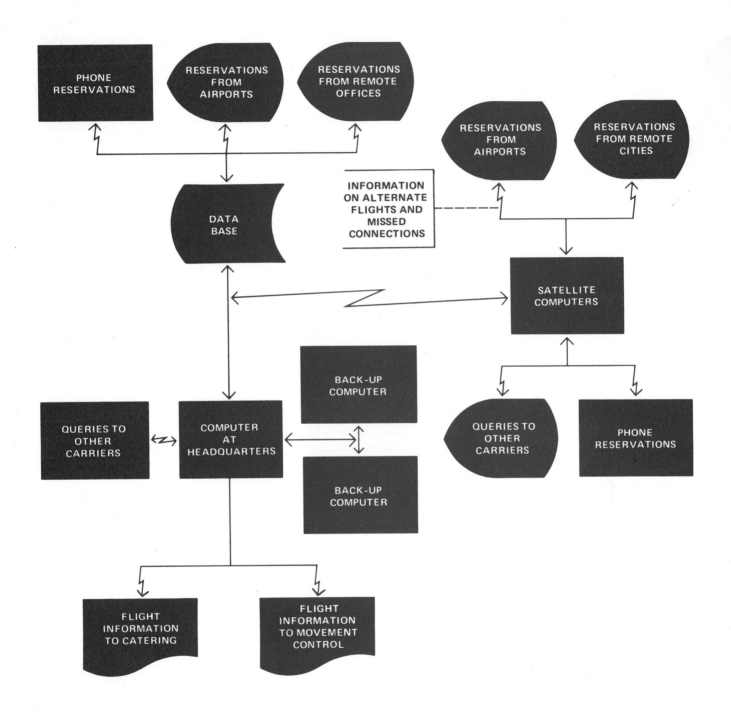

FIGURE 4.4

Chapter Four: Elementary System Flowcharting

4. SKIP COMPUTER PROGRAMMING DETAILS. Indicate the points at which computer programs are to be used for processing and state the operations to be performed. Programming details belong on the program flowchart and should not be shown here.

5. SHOW MAJOR COMMUNICATION LINKS. Relate all parts of the system together, being sure to show communication links. Telephone lines, leased lines, and other links should be indicated.

6. AVOID EXCESSIVE DETAILS. Show only the key elements and avoid less important items. The purpose of the system flowchart is to present a bird's-eye view of the data flow in a system. Too many details or crossed flowlines may cloud the picture.

7. BE NEAT. Use a flowcharting template to ensure uniform, accurate symbols. Don't crowd symbols and flowlines. Adequate margins increase legibility.

Exercises

1. Summarize several fact-finding techniques a systems analyst might use.
2. Describe the goals of systems analysis.
3. Define the term *system implementation*.
4. What activities are carried out in the systems design phase?
5. Draw a flowchart that illustrates the four steps in systems analysis.
6. Study a system (such as a local business or your own school). Write a one-paragraph statement of system goals.
7. Why is systems modification an essential step in systems analysis?
8. Summarize several rules for preparing system flowcharts.
9. Observe a business system and attempt to draw the major elements within the system.
10. Draw an elementary system flowchart illustrating the steps a letter moves through as it is written, mailed, and received by the addressee.

Chapter FIVE
Elementary Program Flowcharting

The steps a programmer follows in structuring the solution to a data processing problem were discussed in Chapter Two. This chapter further explores general programming considerations and specific and applied rules of program flowcharting. Since flowcharting is an integral part of programming, different kinds of flowcharts are presented and illustrated with several sample problems. With an understanding of these concepts and techniques, the student can study the examples of elementary algorithms and applied logic presented in the remainder of the text.

INTEGRATED PROGRAMS AND MODULAR PROGRAMS

The programmer can design a program as a complete, self-contained entity, or as a group of interchangeable modules, each of which performs a different function.

☐ **INTEGRATED PROGRAMS:** An integrated program is composed of many interrelated steps. However, all operations and processes necessary for execution of the program are included. No single sequence of instructions can function independently of the other portions of the program. The program represents the entire problem under consideration. All instructions and processes in an integrated program are written as integral parts of the program, and intermediate results are available for reference by other portions of the program (see Figure 5.1 on page 46).

A change or modification of a process in this type of program often requires one or more corresponding changes in other parts of the program as well. This can lead to errors and complicated testing and debugging.

Integrated programs are usually written by a programmer working alone. The program is designed to handle the specific data input and output and to perform the specific tasks in the most efficient and convenient way.

An integrated program is an efficient and flexible structure for outputting many intermediate results, for making them available to other parts of the program, or for intermixing data input, processing, and output during execution. They are an efficient means of manipulating a large data file in which data records are processed singly. The program might read in a record, process it, output the results, and then loop back (return) to the beginning of the program and execute the same steps on the next piece of data.

☐ **STRUCTURED (MODULAR) PROGRAMS:** In *structured* or *modular programs,* the sequence of instructions that performs each procedure or operation is written as an independent unit, or *module.* This module is capable of performing that procedure in other programs as well.

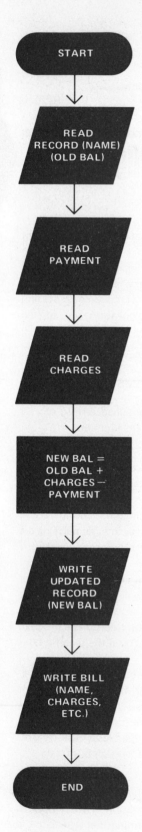

FLOWCHART OF INTEGRATED PROGRAM

FIGURE 5.1

Modular programs are composed of many such independent units tied together by a sequence of instructions called a main program. The *main program* sees that control passes to the proper module in the right sequence and that all the data needed for the module is available. Figure 5.2 is a detail program flowchart of a program with a modular structure.

Each module might correspond to one block on the modular program flowchart. The first module might read in the data needed for processing; the second, perform a sort procedure and rearrange the data into a different order. The third module might perform calculations on the rearranged data; the last, output the manipulated data.

This is a very flexible structure. Processes can be added or deleted by changing only one module and without disturbing the rest of the program. Data can be sorted into a different arrangement by replacing one sort module with another. Or, by removing the calculation module, the data can be rearranged and output in the new order without any mathematical manipulation. A module that performs a specific process can be expanded to handle more exceptions, to allow more alternative paths, or to be more discriminating in the data it processes.

Modular programming offers many advantages:

A. A modular program is more easily debugged, since each module can be removed and tested by itself.
B. Modules can be developed in any order and by more than one programmer.
C. Programming of complex programs is simplified, since several programmers can work on independent modules.
D. These stand-alone units can often be reused in other programs, saving programming time and effort. (Some companies have complete libraries of prepared modules ready for inclusion in new or revised programs.)

A disadvantage of the modular programming structure is the loss of some flexibility of data format design. Since data must be transferred between modules, some degree of standardization is necessary so that all modules will be able to utilize the data.

☐ SUBROUTINES: A closely related form of modular programming is the use of subroutines. In this form, each module, called a *subroutine*, is compiled separately and stored in the computer under a unique program name. The main program calls in the proper subroutine in the correct sequence and transfers control to it. Control returns to the main program after each subroutine performs its function. The next subroutine is then called in. Data to be processed by the various subroutines must be made available by the program at the proper time and in the proper format for each subroutine.

This arrangement has several advantages and disadvantages. Subroutines are usually compiled, tested, debugged, and then stored on secondary storage media as object modules (in machine language). Here they are easily accessible to the calling programs without having to be rekeyed, recompiled, or retested. This saves computer time and expense, as well as programming time and effort.

Processing large programs is often facilitated with subroutines by designing the main program to call only one module into primary computer storage at a time. When that module has been processed, the program reads the next module in and continues processing. Of course, care must be taken to see that data needed by the subroutine is available for access.

Using a subroutine is very advantageous in programs where the same series of steps must be performed several times during execution and where control must return to different points in the program each time. Writing this sequence of instructions as a subroutine and calling it in at the points in the program where it is needed saves recoding many statements. Figure 5.3 illustrates a subroutine used in this manner.

A disadvantage of subroutines is the necessity of standardizing data format for transfer between the program and the modules. This may require additional instructions in the main program to change the structure of the data to make it suitable for a subroutine. And the process may be less unique than a custom-made program.

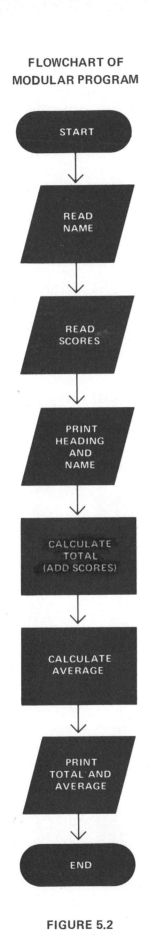

FIGURE 5.2

FIGURE 5.3

Chapter Five: Elementary Program Flowcharting

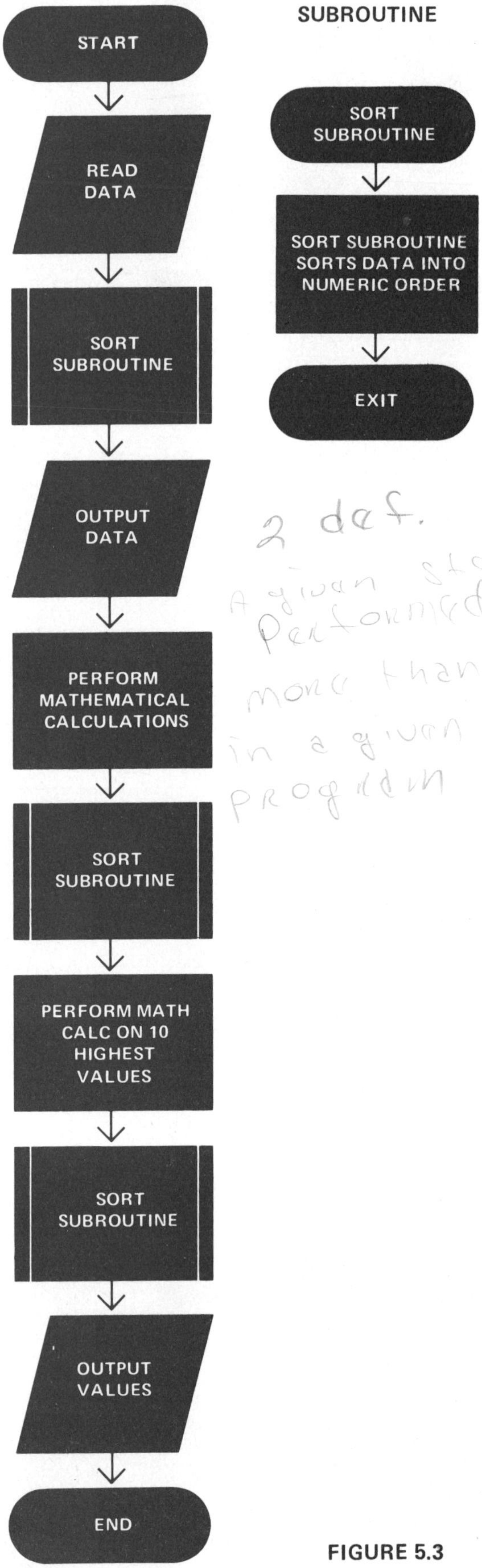

47

SAMPLE DETAIL PROGRAM FLOWCHART

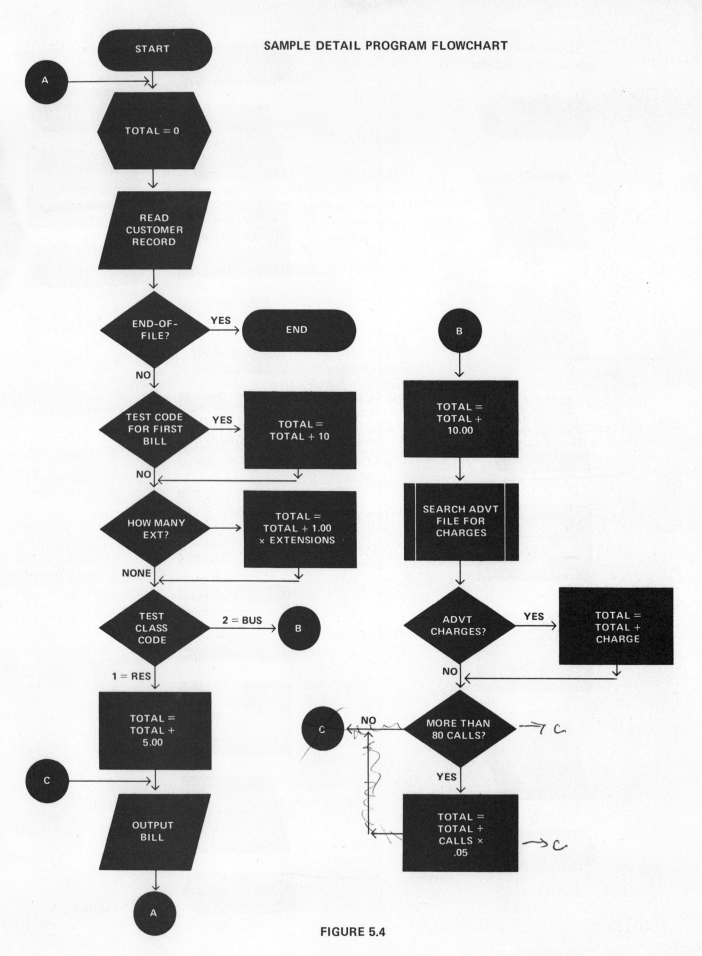

FIGURE 5.4

MODULAR PROGRAM FLOWCHART OF A PAYROLL PROGRAM

Programs can, of course, have both integrated and modular portions. A subroutine that performs a search might be called into an otherwise integrated program (see Figure 5.4) or a program may have input and output functions written in modular form, as well as several integrated arithmetic procedures.

PROGRAM FLOWCHARTS

MODULAR PROGRAM FLOWCHARTS: The modular program flowchart documents the logic flow in the algorithm. It shows the processes performed by the computer program in the sequence in which they are executed. It is sometimes called a *block diagram*.

Figure 5.5 is a modular flowchart of a payroll program. The major steps in the program, such as updating files or calculating taxes and miscellaneous deductions, are shown. Each block may require many computer level steps to be executed in order to perform the process. Figure 5.6 details the striped block. This level of detail is not shown on the modular flowchart.

Modular flowcharts have these advantages:

A. The major blocks, or processes performed, are clearly specified and not obscured by too many details.
B. A bird's-eye view of the whole logic flow shows how each block, or module, is related to the overall logic of the solution.
C. Many potential errors and omissions in the algorithm become apparent from a study of the modular program flowchart.

DETAIL PROGRAM FLOWCHARTS: The detail program flowchart expands the blocks in the modular program flowchart. Each symbol represents an executable step suitable for coding as an instruction in a computer language.

The detailed steps shown in a detail program flowchart may vary slightly when coding in different languages. This is because each language may perform an operation in a slightly different manner or at a different level of sophistication. The major steps in the program logic followed will usually be the same for all languages, however.

In the detail program flowchart, each detail, minute step, or operation is shown in its correct sequence, reduced to its simplest components.

Figure 5.6 shows the "calculate deductions" block from the modular program flowchart as it would appear on a detail program flowchart. Each symbol represents a step that can be coded in one or two instructions in a given computer language.

GENERAL PROGRAMMING CONSIDERATIONS

Several important general considerations which apply to virtually all programming languages affect the preparation of flowcharts. Adherence to the principles below will simplify the process of flowcharting and increase the value of resulting programs.

1. SIMPLICITY. Program logic should be as simple and uncomplicated as possible. Avoid tricky, "cute," or very intricate algorithms.

2. TESTABILITY. Plan the program for maximum testing ease and debugging convenience. Place test points in the program. For example, print out the results of intermediate steps so that they can be checked.

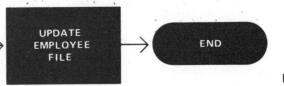

FIGURE 5.5

Chapter Five: Elementary Program Flowcharting

**DETAIL PROGRAM FLOWCHART OF A
MODULE FOR CALCULATING MISCELLANEOUS DEDUCTIONS**

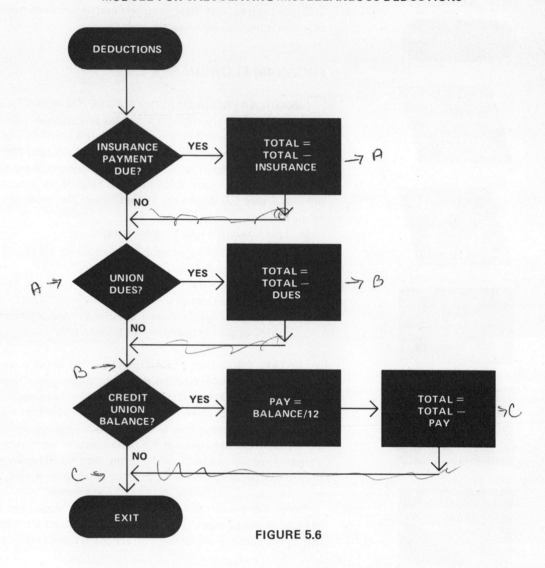

FIGURE 5.6

3. EFFICIENCY. Design the program to use the least amount of primary storage and the fewest devices possible. Avoid reserving excessive storage for tables. Don't place within loops statements that need to be executed only once. Don't read an entire file of data into storage at one time if the data items will be processed singly.

4. MODULARITY. Build programs in blocks or modules. These blocks can be easily modified or expanded later. A library of reusable standardized blocks facilitates preparing other programs or expanding existing ones.

5. TRANSFERABILITY. Plan the program to be as machine independent as possible. Such programs are easier to test and debug and easier for others to follow and use. There will be more documentation and literature available for research and debugging. These programs require less conversion and rewriting efforts if they must be transferred to another system. In one system, special features of operating systems, computer hardware, and language extensions may be great conveniences, while in others they may cause serious problems which are sometimes hard to detect. Follow ANSI, CODASYL (Conference on Data System Languages), or other approved versions of language features.

6. *GENERALITY.* Design the program to be generalized and flexible, if possible. A program that can process varying amounts of data, or different kinds of data, is more valuable than one that accepts only a fixed number of items. Design plots and sorts to handle a varying number of elements. Read titles and headings in as variables instead of as literal text written into the program.

7. *FLEXIBILITY.* Account for all possible varieties of data input and processing conditions. Plan the program to handle unpredictable or incorrect data input, such as zero information, no data, values beyond the given range, alphabetic data in numeric fields, and compensating errors. Include provisions for handling unexpected situations that arise during processing. Have the program print out error messages indicating the specific problem.

DESIGNING GRAPHIC OUTPUT

The layout and format of data output is an important consideration in program design. The value of output is increased if it is easy to read, identified, and complete. Carefully designed output graphics can simplify programming efforts. The following rules will help improve the usefulness and aesthetics of forms, documents, reports, and other output. Figure 5.7 illustrates some of these suggestions.

- A. Start each printout on a new page. This separates the new output data from output of a previously run program.
- B. Break the text at a logical or convenient point when output requires more than one page of printing. Use a line counter or other technique to limit the number of lines on a page. This prevents separating mathematical problems and their answers or printing part of a chart on one page and the rest on the next. This is especially important to remember in dealing with systems that automatically leave some blank lines at the bottom and top of a page.

OUTPUT GRAPHICS

DEPT. 28   PAGE 4

****

WEEKLY SALES BY DEPARTMENT

****

+++ HOUSEWARES +++

SALESPERSON	COMMISSIONED SALES	DISCOUNT SALES	TOTAL PER SALESPERSON
STERLING	$1,098.64	$679.44	$ 1,778.08
MC FADDEN	2,109.13	533.20	2,642.33
ORTEGA	1,782.29	701.80	2,484.09
WEISS	944.63	390.11	1,334.74
JOHNSON	1,703.98	509.61	2,213.59
	TOTAL SALES PER DEPARTMENT -		$10,452.83

****

FIGURE 5.7

C. Label each page with a number and name. A counter can be used to print out a consecutive page number on long programs. It is also a good idea to print out an identifying name at the top of each new page.
D. Label all quantities printed out. Always place descriptive text near printed values for identification. These labels may be handled as variables in a generalized program to increase its flexibility.
E. Line up data. Align totals, subtotals, decimal points, and dollar signs. This helps avoid errors, increases ease of reading, and improves appearance.
F. Use double and triple spacing to separate items. The appearance and legibility of reports and forms can be improved by inserting extra spaces to separate lines of text, figures, and titles.
G. Establish several margins to clarify output. A page is easier to read if different types of data are aligned along different margins. For example, print alphabetic data beginning in Column 1, numeric data in Column 5, and totals in Column 15.
H. Place important lines of text, such as headings or titles, within graphic symbols. Use stars, asterisks, X's, or other symbols to set off important items.
I. Take the built-in features or alignment conventions of some compilers into consideration when designing a program. An example would be the different print fields available in the BASIC language that allow columns to be aligned according to different spacing by changing one character in the PRINT statement.

PREPARING PROGRAM FLOWCHARTS

As stated earlier, many programmers follow the convention of using only certain symbols when drawing modular and detail program flowcharts. These are the basic processing, input/output, flowline and annotation symbols, the decision, predefined, preparation, connector, terminal, and parallel mode symbols. The examples in this text follow this policy.

The rules for drawing flowcharts given in Chapter Two apply to preparing both modular and detail program flowcharts. In addition, the following guidelines may be of help to the programmer in charting a program.

A. In program flowcharting, stress the steps followed in solving the problem, not the media used. The sequence of operations in the algorithms is the important message here.
B. Prepare the modular program flowchart before the detail program flowchart to organize the general structure of the program. This will serve as a guide when you develop the detail program flowchart.
C. Modular program flowcharts should show only the main procedures performed in the program. Each process should be represented by a single symbol.
D. Each procedure, or module, in the modular program flowchart is reduced to programmable steps in the detail program flowchart. Ideally, each symbol on the chart will represent one coded instruction.
E. Show all steps in the algorithm in their proper sequence in the detail program flowchart. All branches and alternate paths of logic flow must be clearly indicated. No branches should be left incomplete.
F. Begin and end each flowchart with the terminal symbols. Exploded or detailed sections removed for expansion should also have these two symbols. Label them in the main program START and END. Label subroutines or expanded sections with a name at the beginning and EXIT at the end.
G. Identify all boxes. Place descriptive text in each block to show precisely what step or procedure is to be taken. On detail program flowcharts, show the action that is performed by each coded instruction. Use annotation symbols to expand this, if necessary.

H. Cross-reference symbols to coded instructions in the program listing for convenience in following the algorithm and error detection.
I. Be consistent. Do not mix the level of modular and detail flowcharting. This helps prevent errors in coding caused by leaving out intermediate steps or incomplete branches.
J. Assign variable names that are as similar to the real names as possible. This makes it easier to identify variables represented on the flowchart. (The names selected will vary according to the language used for coding.)

SAMPLE PROBLEMS

The following example will trace a typical data processing problem through the analysis, decision table, and flowcharting steps.

☐ GREATER METROPOLIS TELEPHONE COMPANY: The Greater Metropolis Telephone Company wants a computer program that will prepare its monthly billing. The bills are to show charges for telephone extensions, service charges, installation, and advertising, wherever appropriate, and should be ready for mailing to customers.

The problem as stated above is too ambiguous to be programmed. Variables are not specified as quantitative terms. Alternatives and conditions are not clearly defined. What are the monthly charges for each type of service? Are business and residential phones billed at the same rate? What are the rates for installation? Is there a limit on the number of calls or a charge for message units?

The first task of the programmer is, therefore, to learn all the information needed to clarify all conditions present. The problem variables must be reduced to definite terms and precise parameters must be set up:

Basic business rate is $10 per month, with a limit of 80 calls. Excess calls are billed at 5¢ per unit.

Basic residential rate is $5 per month, with no limit on calls.

Installation charge for both residential and business phones is $10.

Advertising charges, available from the business department file, are to be added to the monthly bill.

Charge for extension telephones is $1.00 each per month.

To help the programmer visualize and document the relationship of conditions and actions, a decision table is prepared (see Figure 2.3, p. 13). All possible conditions are listed in the condition stub, with all possible combinations shown in the condition entry. Actions are listed in the action stub, in the lower left-hand portion of the table. The action entry portion shows which combination of actions are to be taken for each set of conditions above. In this example there are 20 possible rules (sets of conditions and their actions).

Next, the programmer experiments with various solutions or algorithms with the help of flowcharts. Figures 5.8A and 5.8B show two modular program flowcharts for a monthly billing operation.

Figure 5.8A first tests to see if it is the first bill and how many extensions there are. It then separates clients into residence or business customers before calculating the balance of the charges. Both branches merge and use the same output module. In Figure 5.8B, the records are first separated into residence and business customers and the appropriate charges are calculated. The two branches join to calculate the balance of the charges and to output the bill.

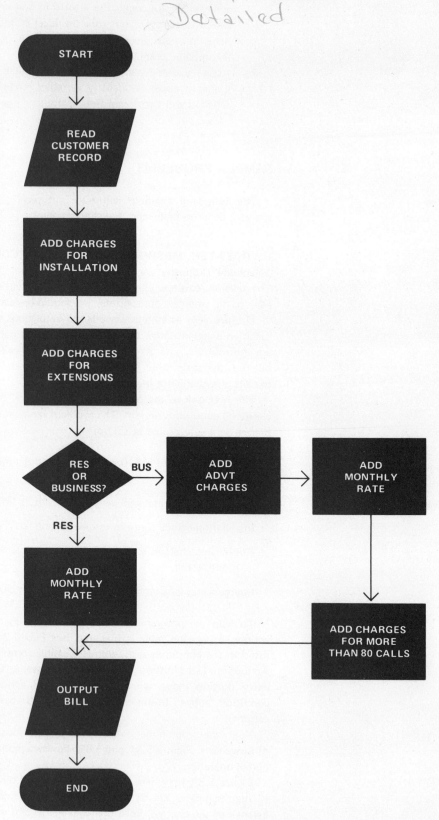

FIGURE 5.8A

MODULAR PROGRAM FLOWCHART B

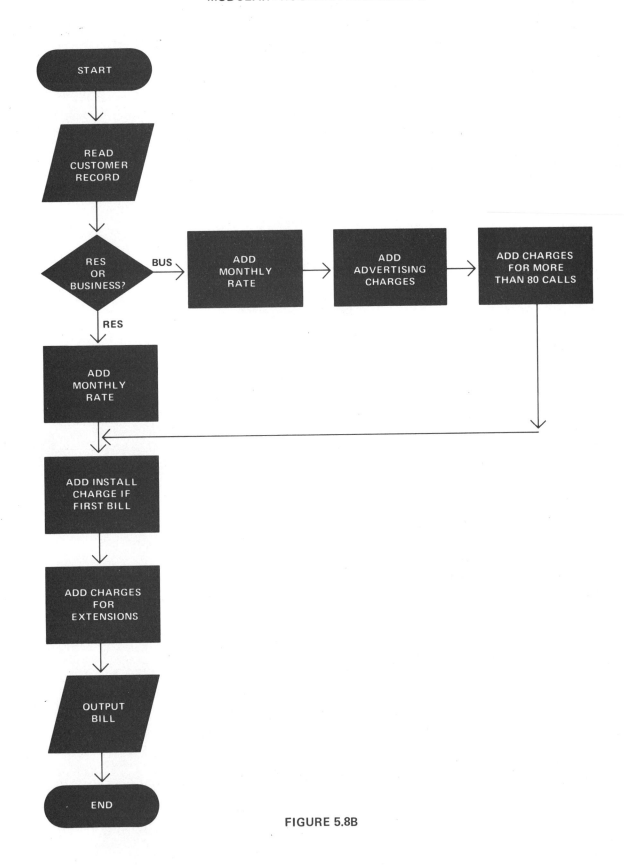

FIGURE 5.8B

Chapter Five: Elementary Program Flowcharting

INVENTORY UPDATE PROGRAM

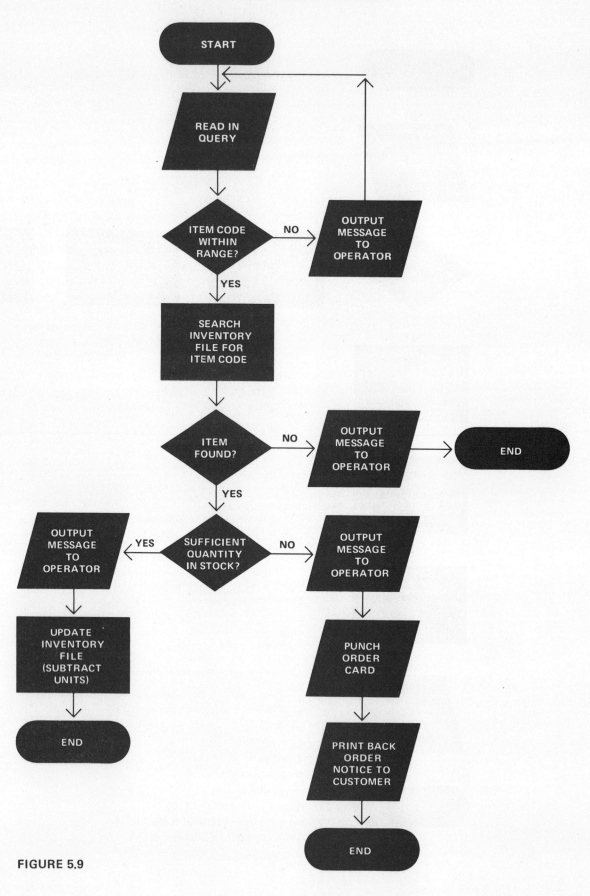

FIGURE 5.9

If the programmer selects the solution in Figure 5.8A and prepares a detail program flowchart, it is assumed that the customer record contains codes indicating whether it is a business or residential phone, whether it is the first bill, how many extensions there are, and how many calls were made.

The blocks in the modular program flowchart are reduced to steps suitable for coding. Note that a subroutine that performs a search is called in to search the advertising file to see if there are charges for the customer. The program calculates the bill for one customer after testing for end-of-file. Then it loops back to prepare the next bill. When the end-of-file is reached, the program terminates execution.

The following examples flowchart computer programs that might be used in the business data processing systems illustrated in Chapter Four.

SALES ORDER SYSTEM. Figure 5.9 is a modular flowchart of a program that checks the inventory file to see if items are in stock in sufficient quantities to fill an order. The program receives a query that lists the item code and number of units needed. It tests the item code to make sure it is within the range of the values stored in the file. If it is not, it outputs an error message to the operator. If it is, it performs a sequential search of the inventory file looking for the item code.

If the code is not found in the file, it outputs a message to the operator and stops execution. If the code is found, the program tests to see if there are sufficient units in stock to fill the order. If there are not, it outputs an appropriate message to the operator, punches out an order card for the item, and prints out a back-order notice for the customer.

If there are sufficient units in stock, the program outputs a message to the operator and updates the inventory file by subtracting the units ordered from the amount in stock.

REAL ESTATE LISTING SYSTEM. Figure 5.10 is a modular program flowchart of a program that prepares the various reports used in this system. It is an example of a program with a modular structure that calls in one of four subroutines.

First the program reads in the code indicating which report is needed and branches to one of four paths. Each branch calls in a different subroutine to prepare a specific report. Subroutine 1 prepares the bimonthly listing and uses the same output module as Subroutine 2, which prepares the daily listing report. Subroutine 3 prepares and outputs the weekly sales report. Subroutine 4 prepares and outputs the geographic report. The program terminates after a subroutine has been executed. To prepare another report, the program must be rerun from the beginning.

Exercises

1. Contrast the differences between modular and detail program flowcharts.
2. Define *integrated programming.*
3. Define *modular programming.*
4. Define a *subroutine.*
5. List five general programming considerations.
6. List several rules that will improve graphic output.
7. Why is it important to identify all boxes?
8. List five rules for drawing program flowcharts.
9. What are the advantages of modular programming?
10. What are the advantages of subroutines?

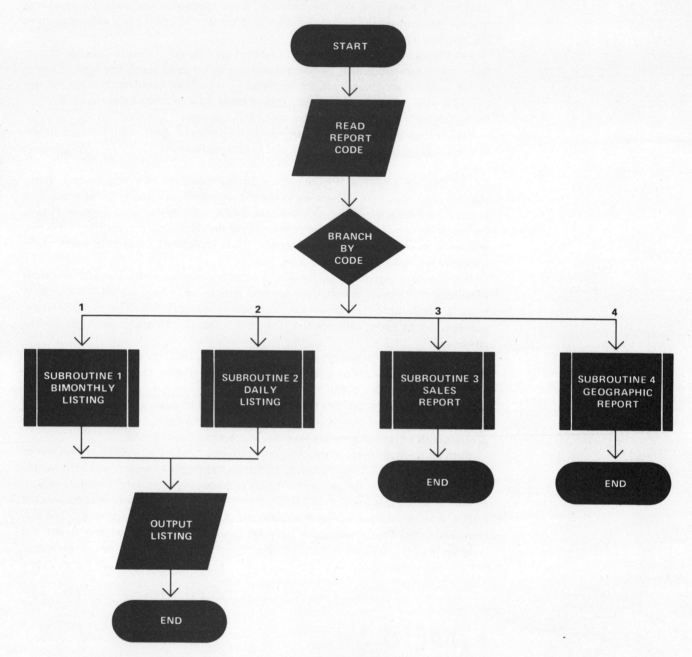

FIGURE 5.10

Chapter Six
Program Building Blocks

This chapter presents examples of elementary program logic. The algorithms are the basic fundamental building blocks that are combined in different ways to create computer programs. With a familiarity of these basic concepts, the student is ready to begin a study of the applied programming logic examples discussed in Chapter Seven.

BUILDING BLOCK 1

Single Pass Execution

FUNCTION • In the single pass algorithm, the computer processes a sequence of instructions one time and terminates execution (see Figure 6.1).

APPLICATIONS • The single pass algorithm is used to execute the processes in a program only once. To repeat the processes on another set of data, it is necessary to reexecute the entire program from START to END. Single pass algorithms might be used to compute a simple-interest problem, to compute a tax bill on one set of data, to print out a block of data from storage, to process a program containing many mathematical procedures on one set of data, or to perform a search.

LOGIC • The computer begins execution by processing the first instruction in the program. Then it executes the next instruction in sequence, then the third, and so on, until the last instruction has been processed. At that point the program execution terminates. If a sequence of instructions must be repeated, the statements must be recoded and placed at the proper points in the program.

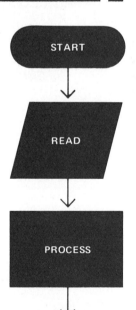

FIGURE 6.1

Exercises

1. Give several examples of problems where a single pass execution might be used.
2. What is the major limitation of a single pass execution?
3. Draw an elementary flowchart of a single pass computation, such as computing simple interest (I=PRT).

Unconditional Branch

FUNCTION • The unconditional branch is used to direct control to a different point in the program rather than to the next sequential instruction.

APPLICATIONS • Unconditional branches are used in algorithms to repeat a sequence of instructions, to repeat the entire program, to skip instructions, and to direct control to an alternate sequence of instructions. Examples would be to process the steps in the program on several different sets of data, to repeat a calculation several times, or to perform alternate processes on data.

UNCONDITIONAL BRANCHES

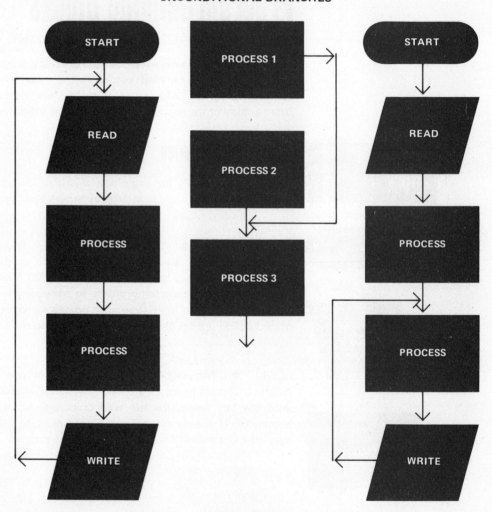

FIGURE 6.2

LOGIC • The computer always executes programming instructions sequentially, from first to last, unless it is directed by the program to do otherwise. The unconditional branch is one method of instructing the computer to stop executing instructions sequentially and to go to a different part in the program to continue processing. The unconditional branch instruction tells the computer where to branch and which statement in the program to go to next.

    The computer processes instructions in order, one after another, until it reaches the unconditional branch instruction. It then goes to the instruction indicated by the branch, executes it, and continues processing the instructions that follow. It does not return to the original branch instruction. The computer continues processing instructions (at the new location) sequentially, until it is directed again to do otherwise.

The GO TO statement is used in FORTRAN, BASIC, and COBOL to perform unconditional branches.

Unconditional branches are usually shown on a flowchart with flowlines or a combination of flowlines and connector symbols.

Exercises

1. How does the unconditional branch differ from the single pass execution?
2. Briefly state the logic followed in the unconditional branch.
3. Under what conditions would an unconditional branch be used?
4. Draw an elementary flowchart showing an unconditional branch.

Conditional Branch (two-way)

FUNCTION • In a two-way conditional branch, the computer executes one of two alternate sequences of instructions, depending on whether or not a test variable meets a specified condition.

APPLICATIONS • Algorithms using conditional branches take advantage of the computer's ability to make relational comparisons and decisions. They can be used to direct the computer to repeat a sequence, to skip a sequence, to separate items into two categories, to perform different processes on data depending upon results of the test, or to select specific data items from a file. Examples would include separation of students into two categories, 21 years and over or under 21 years of age; branching to a closing routine after COUNT=50; branching to an error message if a record is out of order in a file; or branching to a different procedure if a sale is wholesale rather than retail.

LOGIC • A variable is read into a program as a data item, written into a program as part of an instruction, or calculated during processing. An instruction in the program directs the computer to compare this variable against a specified condition. If the condition is met, the computer branches to a certain sequence of instructions and continues processing. If the condition is not met, the computer branches to another set of instructions and continues processing. Branches may also contain instructions that terminate execution of a program.

The specified condition usually involves another quantity or variable and one or two of the relationship signs. The test variable may be compared to see if it is less than, equal to, or greater than another value. The second quantity may be a variable, a constant, or a mathematical expression (in some languages). In many languages the test variable may be included as part of a mathematical expression.

Following are examples of test variables, comparison values, and conditions.

$A*10=30?$ $A=<(C*D)?$ $B+C*4 = A/B?$

$A>B?$ $Y=1?$ $COUNT + 1 = 100?$

Chapter Six: Program Building Blocks

TWO-WAY CONDITIONAL BRANCHES

The test variable does not always have to be the main data item being processed. Another field on a record (the account number, for instance) or a value assigned by a counter or calculated during processing can be used as the basis for testing. The main data item or record will then be processed or manipulated according to the selected sequence of instructions.

In some languages, alphanumeric characters may also be compared. In these cases sequencing is usually done in alphabetic order.

The test variable and condition are often phrased on the flowchart symbol to produce a YES or NO, TRUE or FALSE answer. This facilitates following logic flow.

In FORTRAN, BASIC, and COBOL languages, the IF statement is used to direct the computer to perform testing and branching.

Coding Note: Each pathway of a conditional branch leads to a different set of operations. When coding, each set of operations is converted into a group of instructions. These groups are usually placed one after the other in the written program. Therefore, an unconditional branch must be the last statement in each group to route control around the other groups and send it to the next appropriate point. Otherwise, the program will continue to process statements sequentially and execute all the groups instead of only one.

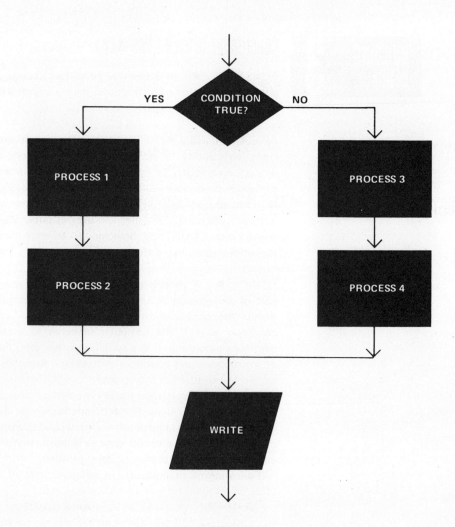

FIGURE 6.3

Computer Algorithms and Flowcharting

Exercises

1. Briefly state the logic followed in the two-way conditional branch.
2. List several reasons why a conditional branch would be used in a program.
3. How does a conditional branch differ from an unconditional branch?
4. Flowchart a conditional branch that repeats a sequence with three steps.
5. Flowchart a conditional branch with two parallel paths.
6. Flowchart a conditional branch in which one branch performs two steps and then rejoins the main program flow.

Conditional Branch (three-way)

FUNCTION • In the three-way conditional branch, the computer executes one of three alternative sequences of instructions, depending on the relationship of a test variable to a specified condition.

APPLICATIONS • Algorithms with three-way conditional branches are used to sort items into one of three categories, depending on whether a test variable is less than, equal to, or greater than a test condition. Each condition branches the computer to a different sequence of instructions. This greatly increases the power and flexibility of computer programs. A program can branch to different processing routines depending on whether one price is less than, equal to, or greater than another. A program can test the number of parts in stock against the number 50 and process each category in an appropriate manner. Payments can be checked to see if they are prepaid, on time, or past due and handled accordingly. The idea is to have the program test all the alternatives possible and then to provide a different set of operations for each of them.

LOGIC • The value of a test variable is read or written into a program or calculated by it. An instruction in the program directs the computer to compare this variable against a specified condition. If the variable is equal to the condition, the program branches to a certain sequence of instructions and continues processing. The program branches to a different sequence of instructions if the variable is greater than the condition and to yet another if the variable is less than it. Branches may contain instructions that terminate program execution or direct control to another part of the program.

In most languages, the test condition may be a variable, a constant, or a mathematical expression. The test variable may be used in a mathematical expression in many languages.

Here are some examples:

Is CODE =, >, or < 2000?   Is TEMP =, >, or < 0?

Is A 1 =, >, or < C*.298?   Is XXX+3 =, >, or < YYY?

Alphanumeric characters may be tested in some languages. They will usually be sequenced in alphabetic order.

The test variable does not have to be the main data item being processed. It can be another field on a record (account number, number in stock, or classification code) or a value related to the data item and calculated during program execution (counter or mathematical result).

Not all languages contain statements that allow direct coding of three-way branches. In these cases, two two-way branches may be required to achieve the same differentiation when processing sequences. For example, the branch instruction in FORTRAN, BASIC, and COBOL is the IF statement. In FORTRAN and COBOL, three-way branches are possible with one statement. In BASIC, two two-way branches would be necessary.

Test conditions and alternate branches should be carefully labeled on flowchart symbols to facilitate following program logic flow.

Coding Note: Each pathway of a conditional branch leads to a different series of operations. When coding, each of these series is converted into a group of instructions. These groups are usually placed one after another in the written program. An unconditional branch must be the last statement in each group to route control around the other groups and send it to the next appropriate point. Otherwise, the program will continue to process statements sequentially and execute all the groups, instead of only one.

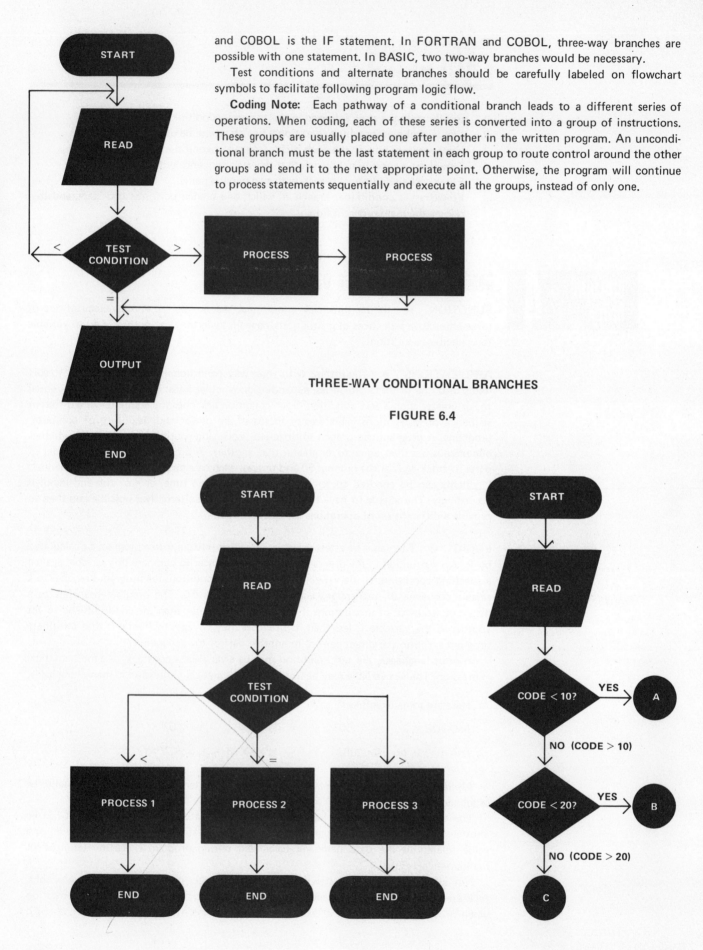

THREE-WAY CONDITIONAL BRANCHES

FIGURE 6.4

Computer Algorithms and Flowcharting

Exercises

1. Briefly state the logic followed in the three-way conditional branch.
2. List several examples showing where a three-way conditional branch would be used in a program.
3. Why must an unconditional branch be the last instruction in each pathway of a conditional branch?
4. Flowchart a three-way conditional branch that repeats an input sequence or performs two parallel processes.

Conditional Branch (multiway)

FUNCTION ● In the multiway conditional branch, the computer executes one of several possible paths, depending upon the relationship of the value tested to a specified condition.

APPLICATIONS ● Algorithms with multiway conditional branches are used to sort or categorize data into several groups or to select one of several possible routines to perform. This is done by comparing the test variable against several test conditions in sequence until the appropriate group or classification for the variable is found.

A program that categorizes a group of data by decade may have a sequence of test conditions that test a value to see if it is equal to or less than 1910, 1920, 1930, and so on, to 1970. A test variable of 1925 would direct the program to branch to the appropriate routine when it executes the instructions that test for equal to or less than 1930.

A program may also test a code number and branch to the appropriate sequence of instructions. The test conditions in this situation would test to see if the variable is equal to 1, 2, 3, and so on.

Multiway conditional branches can be used to group cash register transactions by department (produce, meat, delicatessen, or dairy). They can classify students into age groups. Such algorithms could also be used to call in a different subroutine for each category of items or to print out a different message depending on the grade a student earned.

LOGIC ● There are several ways to program a multiple-condition branch. The method selected will depend on the nature of the data being tested and the options of the language used for coding. The most common way to program multiple branches is with a series of two- or three-way branches. The value of the test variable is read or written into the program or calculated by it. The first decision point tests the variable against a specified condition. If the variable is equal to the condition, the program branches to the appropriate sequence of instructions and continues processing. If the variable does not equal the test condition, the program is directed to the next decision point. Here the variable is tested against the next condition and the program branches if the condition is met. If it is not, the program moves to the next decision point. This process continues until the appropriate condition for that variable is reached and the proper branch taken. Programs of this type often include conditions that test for errors, such as a value above or below an acceptable range, or a fraction where a whole number should be.

Test conditions are usually composed of one or more variables or constants, or a mathematical expression, and one or two of the relational operators ($>\geqslant<\leqslant=\neq$). In some languages, the test variable may be part of a mathematical expression.

Here are some examples of valid test variables and conditions:

YEAR ≥ 1940?   COLOR = 3?   X + 10 = T/Y?
AGE > 50?   F = D*6?

The test variable and condition are often phrased on the flowchart symbol to produce a YES or NO, TRUE or FALSE answer. All conditions and paths should be carefully labeled to facilitate following program flow. In FORTRAN, BASIC, and COBOL, one or more IF statements are used for multiple-condition branching.

Another way to program multiple branches is to take advantage of a convenient option present in several languages, which permits multiple branching using only one instruction. This option directs the computer to test the variable to see if it is equal to 1, or 2, or 3, and so on. The test numbers must be chronological and begin with 1.

To use this option, the programmer sets up a code system to represent the different categories of data. For example, the decade between 1900 and 1910 might be assigned a code value of 1; 1911 to 1920, 2; 1921 to 1930, 3; and so on. The test value may be a variable or a mathematical expression, depending on the language.

Coding Note: Each pathway of a conditional branch leads to a different set of operations. When coding, each set of operations is converted into a group of instructions. These groups are usually placed one after another in the written program. An unconditional branch must be the last statement in each group to route control around the other groups and send it to the next appropriate point. Otherwise, the program will continue to process statements sequentially and execute all the groups of instructions, instead of only one.

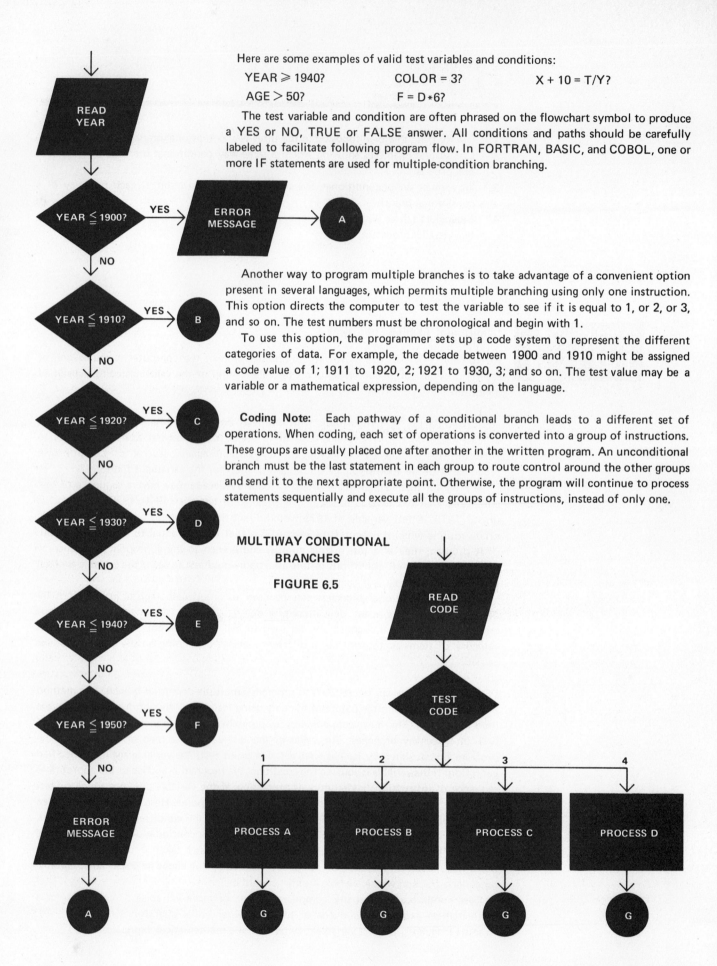

MULTIWAY CONDITIONAL BRANCHES

FIGURE 6.5

66

Computer Algorithms and Flowcharting

Exercises

1. Briefly state the logic followed in a multiway conditional branch.
2. List several applications for a multiway conditional branch.
3. How do multiway branches differ from two-way branches?
4. Flowchart a multiway branch with five branches.
5. Flowchart a multiway branch that sorts data into four categories, depending on whether a value is between $1.50 and $2.50, $2.51 and $3.50, $3.51 and $4.50, or $4.51 and $5.50.

Simple Loop

FUNCTION • A loop directs the computer to return to the beginning of a sequence of instructions and to execute them another time. Statements within the range of the loop will be reexecuted each time it is repeated.

APPLICATIONS • Loops are one of the most useful tools of the programmer. They are able to direct the computer to repeat a process many times without the necessity of recoding the steps to be reexecuted. A loop that includes all of the statements in a program within its range can be used to repeat the processes in the program on another set of data. Loops can be used to repeat one or more of the processes in a program. The input instructions in a program can be repeated 100 times to read in 100 different names. A letter containing 25 different lines of text can be prepared with only one set of output instructions, repeated 25 times within a loop.

An algorithm that calculates and prints out the average of a group of numbers uses a loop. The loop includes within its range the statements that read in a value and add it to a running total. Outside the range are the statements that divide the total by the number of items added as well as the sequence that prepares the output.

LOGIC • A simple loop is a sequence of instructions with a branch as the last instruction. The computer executes the instructions in the loop one at a time, sequentially, until it reaches the last instruction. If this is an unconditional branch, control returns back to the beginning of the loop for another iteration of the sequence. The computer again executes all of the instructions in the sequence until it reaches the last instruction. Again, it branches back to the beginning of the loop to reexecute the statements. This repetition continues until the program directs the computer to terminate the looping.

The most common method for terminating loops is a conditional branch located within the range of the loop. A branch instruction at the end of the loop can either branch control back to repeat the loop or branch to the next process, depending upon whether or not a certain condition has been met. An unconditional branch may be placed in the middle of the loop to compare the number of repetitions against a predetermined amount or test for end-of-file. Methods for terminating loops are explained in more detail in Building Blocks 10, 11, and 12.

In FORTRAN, the GO TO and DO statements are used to program simple loops. GO TO and FOR/NEXT statements are used in BASIC, and GO TO and PERFORM statements are used with COBOL.

Looping is shown on flowcharts by flowlines or by a combination of flowlines and connector symbols.

SIMPLE LOOPS

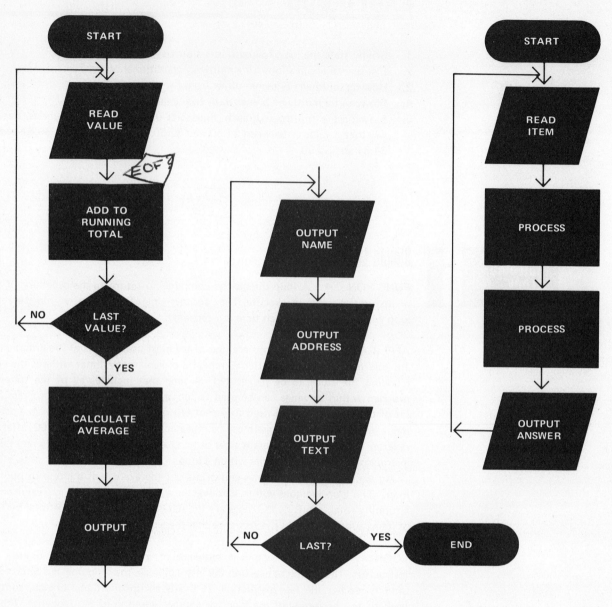

FIGURE 6.6

Exercises

1. Briefly state the logic followed in a simple loop.
2. When are loops used in a program? Give several examples.
3. Describe what happens when the last statement in a simple loop is an unconditional branch. What happens when it is a conditional branch?
4. Flowchart the two loops described in Exercise 3.

Counters

FUNCTION ● Using algorithms with counters is a programming technique that enables the programmer to count or limit the number of times a process is performed. The value of the counter can be used as a test variable in branching, as a variable in mathematical operations, and as a value when indexing or subscripting.

APPLICATIONS ● The use of counters is a valuable and versatile coding technique. Counters can perform many functions in a program. They can be used to count the number of times a process is performed. This allows the programmer to use different sized data sets in a program that needs to know the number of items or records being processed. A program with several alternate branches may use counters to total and print out the number of times each alternate path has been executed. For example, the program might print out that "6 students received As, 15 students received Bs, 40 students received Cs."

Counters are also used to limit the number of times a process is repeated. For example, the program may count the number of data items read in and use this tally to limit the number of times the output instructions are executed. Counters can be used to give consecutive numbers to output forms or pages, and the tally from counters is often used as part of a mathematical calculation (as in the case of figuring the average of a list of numbers).

LOGIC ● Counters are composed of two parts—an initializing statement and an incrementing statement. The first sets the beginning or initial value of the counter. The second is located within a loop and indicates the increment or how much the value of the counter should change each time it is executed. The computer will execute the initializing statement first and assign the value shown to the counter. It continues to process instructions as directed by the program. When it reaches the incrementing statement, it changes the value of the counter by the amount indicated in the statement. Since the incrementing statement is within a loop, it will be reexecuted each time the instructions in the loop are repeated. The computer will add the new amount to the old value of the counter each time it is executed. This process will continue until the loop ends or until the counter is reinitialized.

Counters may increase or decrease each time they are executed. In most languages the computer is instructed to add the new value of the counter to the old by placing the name of the counter on both sides of the equals sign:

    COUNT=COUNT+1   PAGE=PAGE+1

The increment may be a constant, a variable, or a mathematical expression. Examples of counters are:

NO=100	I=0	CT=60	K=0
NO=NO - 5	I=I+N	CT=CT+(X*2)	K=K+2

After a loop ends, the final value, or tally, of a counter remains unchanged until it is changed by another instruction or reinitialized, or until the loop is reentered. This tally can be accessed at another point in the program by reference to the counter's name. It can be used as a variable in another programming statement or in a mathematical equation or to end a loop.

Here is how a counter is used to end a loop: each time the loop is repeated, the counter is incremented and then tested by a conditional branch in the loop. If its value does not equal the test condition, the loop repeats and the counter increments and is tested again. When it finally does equal the test condition, the program leaves the loop and branches to another point in the program. The counter and conditional branch can, of course, be set up to terminate the loop only when the counter does not equal or is greater than or less than the test condition.

The test condition in the conditional branch that ends a loop must be carefully selected. The programmer must make sure that the test condition will be met during

Chapter Six: Program Building Blocks

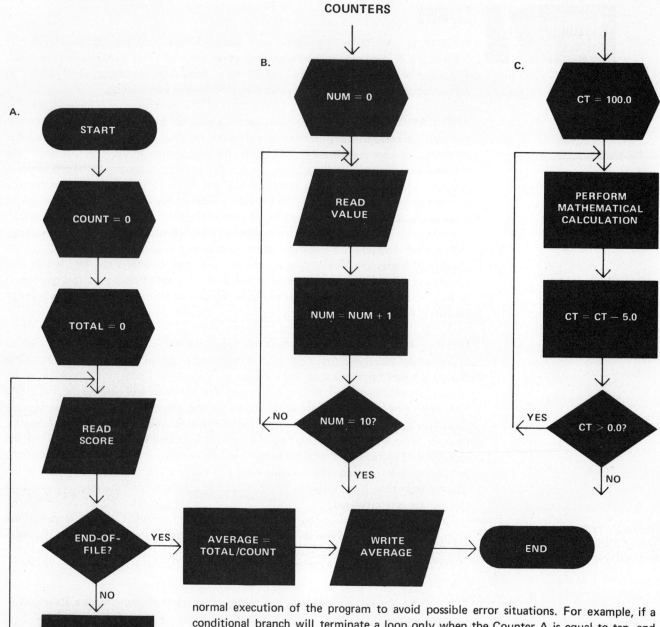

FIGURE 6.7

normal execution of the program to avoid possible error situations. For example, if a conditional branch will terminate a loop only when the Counter A is equal to ten, and through an error in programming, A is accidentally set to 11, an endless loop will be initiated. Situations such as these can often be avoided with the use of the relational operators greater than ($>$) or less than ($<$) instead of the equal to ($=$) condition. This technique is especially important when the increment is a value other than one.

Counters are programmed using the assignment statement in FORTRAN and the LET statement in BASIC. In COBOL, the VALUE clause, the MOVE statement, and arithmetic statements are used to program counters.

Figure 6.7A flowcharts a program with two counters. Both are initialized to 0. One has an increment of 1; the other increment is a variable. Tallies of both counters are used in a mathematical calculation later in the program. The program calculates the average of a group of scores. The first counter, COUNT, tallies the number of scores read in. The second, TOTAL, increases by the value of SCORE each time the loop is repeated. When the end-of-file is reached, the program branches to calculate the average. The value of TOTAL is divided by the value of COUNT to produce AVERAGE.

Figure 6.7B flowcharts a counter that is used to limit the number of times a loop is repeated. The value of the counter is initialized to zero and increases by one each time the loop is executed. A conditional branch at the end of the loop tests the value of the counter and repeats the loop until it is equal to ten.

Figure 6.7C flowcharts a negative counter that decreases by five each time the loop is repeated. A conditional branch at the end of the loop tests the value of the counter to see if it is greater than zero. If it is, the loop is repeated. The counter decreases by 5 and is tested again. When the counter is finally less than or equal to zero, the program "falls through" to the next instruction in the program (which would be represented by a process block below the decision block).

Figures 6.8A, 6.8B, and 6.8C are all examples of incorrect placement of counters.

INCORRECT PLACEMENT OF COUNTERS

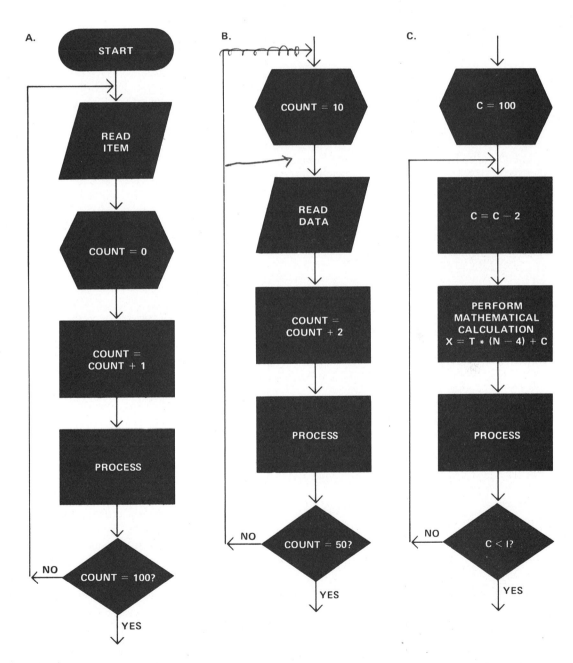

FIGURE 6.8

Chapter Six: Program Building Blocks

Placement of initializing and incrementing statements in an algorithm is of critical importance and should be carefully studied to avoid errors.

Both Figures 6.8A and 6.8B are examples of positive counters. The initializing statements have been included within the loop and will be reinitialized each time the loop is executed. The value of COUNT in Figure 6.8A will never be higher than 1. In Figure 6.8B, COUNT will never be more than 12.

Figure 6.8C is an example of a negative counter. In Figure 6.8C, the decrementing statement is placed at the beginning of the loop. C will be equal to 98 the first time the mathematical equation is calculated and equal to 0 the last time through. The equation will never be calculated with C equal to 100. If the decrementing statement had been placed after the processing symbol, C would have been equal to 100 the first time the mathematical equation was calculated and equal to 2 the last time. The equation would never have been calculated with C equal to 0.

The use of counters in indexing and subscripting is explained in Building Blocks 13, 14, and 15.

Exercises

1. Briefly state the logic followed in flowcharting a counter.
2. What are counters used for?
3. What type of logic errors could occur because of incorrect placement of counters?
4. Flowchart a program segment which has a counter that is initialized to 10 and decreases by 5 each pass.
5. Flowchart a program segment with a counter and conditional branch.

Sequential Loops

FUNCTION • Sequential loops direct the computer to perform two or more simple loops in sequence. The computer executes the instructions in the first loop the specified number of times and proceeds to execute the instructions of subsequent loops. Each loop will usually perform a different procedure.

APPLICATIONS • Sequential loops are usually used in computer programs to manipulate a file, or group of items, as a whole. The instructions in a loop will be repeated on each piece of data in the group. Then the program moves to the next loop; all instructions in that loop will be executed on the data items in the group. For example, the instructions in the first loop might read in a list of numbers. Those in the second loop might perform a calculation on each number, and a third loop might be used to output the results one number at a time.

Sequential loops can read in a list of costs, calculate markup percentage and selling price for each cost, and print out a list showing each cost, markup percentage, and selling price. A sequential loop program that merges a master file with several detail files to produce a new master file could operate in this manner: the first loop reads in the master

SEQUENTIAL LOOPS

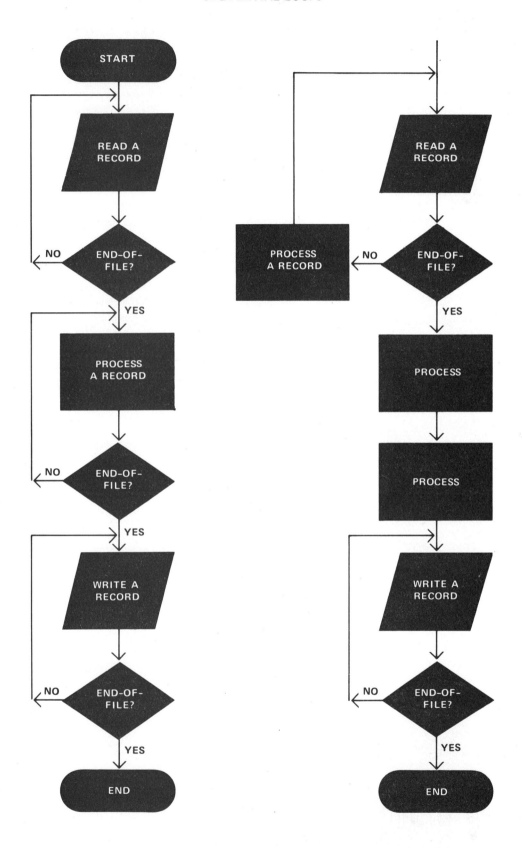

FIGURE 6.9

Chapter Six: Program Building Blocks

file. The second and third loops read in the two detail files. The fourth loop merges the records, and the last loop outputs the merged file.

One or more of the loops in an algorithm with sequential loops can also be used to repeat a series of instructions on a single variable.

Sequential loops save considerable recoding time and effort for the programmer as well as computer compilation and execution time.

LOGIC ● Each loop in a sequential loop program is a simple loop—a sequence of instructions with a branch as its last instruction and some method of termination. The computer executes a program with sequential loops in the following way.

Instructions in the program are processed sequentially until the branch at the end of the first loop is reached. If the branch is unconditional, the program immediately returns to the statement indicated by the branch instruction, which will be the beginning of the loop. The sequence of instructions will contain some form of test that terminates the looping at the proper time and sends control to the next instruction outside the loop. The methods of terminating loops are explained in detail in Building Blocks 10, 11, and 12.

If the branch at the end of the loop is a conditional branch, the computer performs the relational test indicated and either returns to the beginning of the loop to repeat the sequence of instructions or moves to another point in the program.

In any case, when a loop has been executed the number of times directed by the program, control moves to the next instruction outside the loop. In the program with sequential loops, this may be the first instruction in the second loop. The computer will execute the group of instructions sequentially until it reaches the branch at the end of the second loop. As before, it repeats the loop until the conditions of the conditional branch (either at the end or in the middle of the loop) have been met. At this time, it moves to the next instruction.

This process continues until the computer has processed each loop in turn and reached the end of the program.

On flowcharts, sequential loops are indicated by flowlines or a combination of flowlines and connector symbols.

The DO or GO TO statement is used in FORTRAN to perform sequential loops. In BASIC, the FOR/NEXT and GO TO statements are used. COBOL employs the GO TO and PERFORM statements.

Exercises

1. Briefly state the logic followed in sequential looping.
2. When are sequential loops used?
3. Flowchart a program with three sequential loops.
4. Flowchart a program with two sequential loops. Include a counter in one of the loops.
5. Flowchart a program with two sequential loops that reads in a file, performs three processes on it, and prints out the records in the file.

Nested Loops

FUNCTION ● Nested loops are formed when one loop is placed inside the range of another loop. They direct the computer to execute the processes in the inner loop several times for each time the outer loop is executed.

APPLICATIONS • Nested loops allow the programmer to perform one or more repetitive operations on one data item or record and repeat these operations on each data item until the entire file or group has been processed. The inner loops may be sequential or nested. The number of repetitions performed on each data item may vary.

One outer loop and two sequential inner loops may be used in processing checking account balances (see Figure 6.12A). The outer loop reads in a customer's name, account number, and old balance. The first inner loop reads and processes deposits for that account; the second inner loop reads in and processes the withdrawals. Control returns to the outer loop to prepare a master record with the new balance and then branches back to repeat the processes on the next account.

Three nested loops could be used to prepare a grade report for each student in a group (see Figure 6.12B). The outer loop reads in a student's name. The middle loop reads in a class name. The inner loop reads in all the grades earned by the student in that class and adds them to a running total. The middle loop takes control again and calculates and prints out the final grade for that class. Control returns to the top of the middle loop to calculate the grade for the next class. After the grade for the last class has been printed out, control returns to the outer loop to repeat the process for the next student.

Nested loops are used in many types of mathematical problems and data manipulating routines, such as sorts and searches.

LOGIC • In processing a program with nested loops, the computer executes statements sequentially until it is directed by a branch instruction to move to another point in the program. For example, a computer could execute a program with two sequential loops nested within an outer loop, as shown in Figure 6.12A. Assume that in this case there are three customers, with four deposits and five withdrawals each. Therefore the outer loop is to be repeated three times. The first inner loop is to be repeated four times and the second inner loop five times for each execution of the outer loop. The computer, processing statements sequentially, will enter the outer loop and reach the conditional branch. Since it has not yet read the end-of-file, it enters the first inner loop. It continues to process statements sequentially until it reaches the conditional branch at the bottom of the loop. This completes one repetition of the first inner loop. Since the fourth deposit has not been read, control returns to the top of the inner loop to repeat the instructions for a second, third, and fourth time. When the last deposit has been read in, the conditions for the branch are satisfied, and the computer moves to the next inner loop.

It continues to process instructions until it reaches the branch at the bottom of the second loop. This completes one repetition of the second inner loop. The computer returns to the beginning of the second inner loop and executes it a second, third, fourth, and fifth time. The last withdrawal has now been read and processed, and the conditions for leaving the loop have been met.

Control returns to the outer loop. In some programs, the computer may be directed to perform one or more operations at this point. The first execution of the outer loop is completed when control reaches the terminating branch. Control returns to the beginning of the outer loop to repeat the entire cycle for the next account. In this example, the outer loop will be repeated a total of three times, the first inner loop, a total of 12 times, and the second inner loop, a total of 15 times.

Here is how a computer executes a program with three nested loops. Assume that in the example in Figure 6.12B there are four students with two classes each and five grades per class. Therefore the outer loop is to be executed four times, the middle loop two times, and the inner loop five times. The computer begins processing the instructions sequentially, enters the outer loop, and reaches the conditional branch. Since the end-of-file has not been read, control continues into the middle loop and then into the inner loop. When it reaches the conditional branch, it has completed the first execution of the inner loop. Since the last grade has not been read, control returns to the beginning of the inner loop to reexecute it a second, third, fourth, and fifth time. When the last grade has been read in, the conditions for leaving the loop have been met, and control returns to the middle loop. The computer will continue processing instructions until it reaches the terminating branch of the middle loop. The computer has now finished the first execution of the middle loop. Control returns to the beginning of the middle loop for the second execution. The inner loop will be executed five more times and control will return to

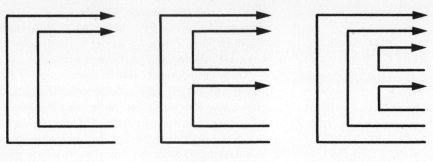

FIGURE 6.10

complete the middle loop a second time. The computer has now completed the required number of repetitions of the middle loop and control returns to the outer loop.

In this example, the conditional branch at the end of the middle loop is also the terminating branch for the outer loop. The computer has completed one execution of the outer loop at this point and returns to the beginning of the outer loop to repeat the cycle. Again the middle loop will be executed one time and the inner loop five times; then the middle loop is executed once again and the inner loop another five times. This will complete the second execution of the outer loop.

In this example, the entire cycle will be repeated two more times before the program terminates. The outer loop will be executed a total of four times, the middle loop a total of eight times, and the inner loop a total of 40 times.

Nested loops are shown on flowcharts by using flowlines or a combination of flowlines and connector symbols. To avoid confusion between flowlines of nested loops, some programmers use dotted lines for inner loops and solid lines for outer loops.

In some languages, the number of loops that may be nested is limited to three. Most languages do not permit branching into a loop except at its beginning. Most languages do permit branching out of loops at any point. Figure 6.10 illustrates some acceptable arrangements of nested loops. Figure 6.11 shows some unacceptable nested loop arrangements.

The DO or GO TO statement is used in FORTRAN to perform nested loops. In BASIC, the FOR/NEXT and GO TO statements are used, and COBOL uses the GO TO and PERFORM statements.

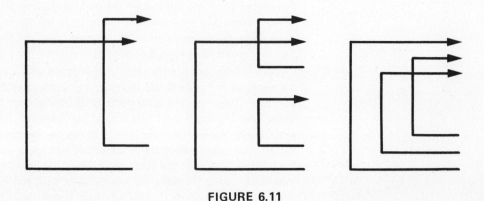

FIGURE 6.11

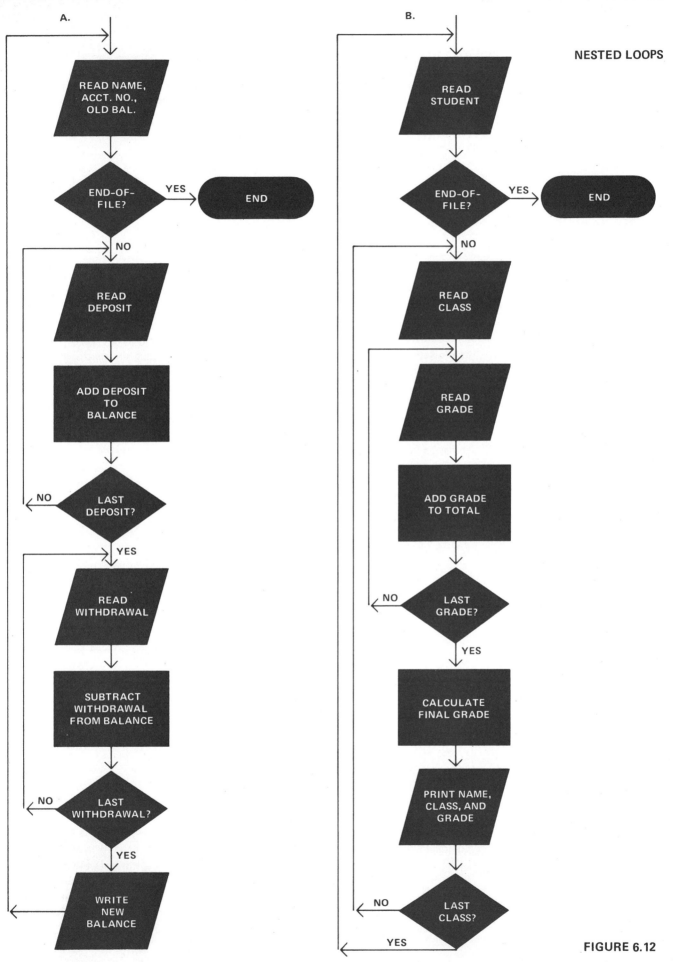

NESTED LOOPS

FIGURE 6.12

Chapter Six: Program Building Blocks

Exercises

1. Briefly state the logic followed in programming nested loops.
2. Give several applications for nested loops.
3. Draw a flowchart with three nested loops.
4. Draw a flowchart with two nested loops. Include a counter in the outer loop.
5. Draw a flowchart with two sequential loops nested within an outer loop. Include a counter in each inner loop.

Terminating a Loop – The Trailer Record

FUNCTION • This technique directs the computer to test each record processed by a loop to see if it contains a specific value in a certain field. If it does, the loop is terminated. If not, the loop is repeated. The record containing the specified value (the *trailer record*) indicates that the end-of-file has been reached.

APPLICATIONS • A trailer record tells the computer that it has processed all the records in a file and instructs it to move to the next process. It is a convenient and practical way of designing a loop to handle an undetermined number of data records. Trailer records are used to limit loops that read in data, process data mathematically, and sort, merge, search, and output data.

Trailer records are used in two ways to terminate processing of a loop. In one method, one field of a record is reserved exclusively for the trailer value. Only the trailer record has data in that field. It is blank on all other records in the file.

For example, if punched cards are being used as the input medium, columns 78-80 may be reserved for the trailer value. The programmer will place a card that contains the value 999 punched into columns 78-80 as the last card in that data set. The program will read in a record and test to see if it has 999 in those columns. If it does not, the card is processed and the next card is read in. If the card read in does contain 999 in columns 78-80, the loop terminates. This method is illustrated below.

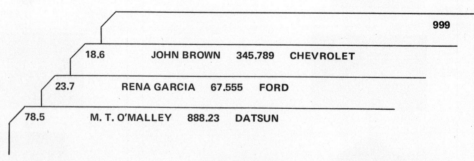

FIGURE 6.13

The second way to terminate a loop with a trailer record is often used in interactive programming. The program is designed to test each data item entered to see if it is the trailer value, such as 99999. If it is not, processing continues and the next item is read in. If it is 99999, the loop terminates. In this instance, the same field is used to contain both the data item for processing and the trailer value. Care must be taken to see that the trailer value would not normally occur in the data range. This method is illustrated in Figure 6.14 on page 79. In some languages, alphabetic values may be used as trailer values.

78.5	M. T. O'MALLEY	888.23
23.7	RENA GARCIA	67.555
18.6	JOHN BROWN	345.789
99999	ZZZ	99999

FIGURE 6.14

LOGIC ● Terminating a loop with a trailer record utilizes a conditional branch to test a data field and branch to one of two possible paths, depending on whether or not the condition is met.

A loop that reads in a data file could be terminated using the following method. The computer reads in a data record containing data in one or more fields. A conditional branch checks to see if there is data in the specified field and compares it to the test condition. If the condition is not met, the loop is continued or repeated. If it is met, control branches to another part of the program. This is illustrated in Figures 6.15A and 6.15B. Figure 6.15C shows a trailer being used to terminate a loop which includes both a process and an output sequence.

Placement of a conditional branch in a loop requires careful consideration. Figure 6.15D shows a logic error resulting from incorrect placement of the branch. The trailer value 99999 will be added to the running total before control leaves the loop, giving an incorrect total. Any further processing that uses this total will also give incorrect results.

Special care must be taken when counters are used in loops terminated by a trailer value. In Figure 6.15E (on page 80) the trailer record (containing ZZZ in the name field) will be counted as a processed record, creating an erroneous count.

Trailer values may be composed of one or more numeric digits. Alphanumeric test values may also be used in some languages. The test condition and the trailer value must be exactly the same; if the conditional branch is looking for 99999, a trailer value of 9999 will not work.

TERMINATING A LOOP WITH A TRAILER RECORD

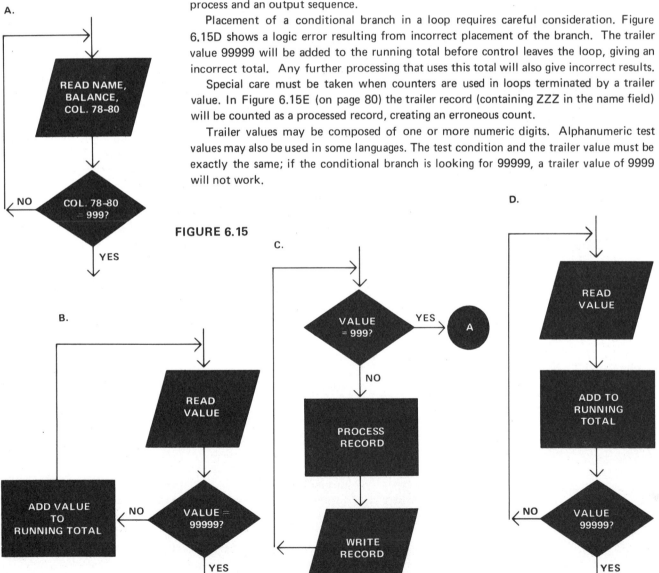

FIGURE 6.15

E.

FIGURE 6.15 (CONTINUED)

Terminating a loop with a trailer is shown on a flowchart by a conditional branch. The location of the branch, the field tested, the test condition, and any alternate paths should be indicated.

The IF statement is used in FORTRAN, BASIC, and COBOL to terminate loops with trailer values.

Exercises

1. Briefly state the logic followed in terminating a loop with a trailer record.
2. What is a trailer record?
3. Where are trailer records used? What are the advantages of using trailer records?
4. Flowchart a program module that reads in a data file. Terminate the loop by testing for a trailer value in an unused field.
5. Flowchart a program module that reads in a data file of records containing three numbers each. Test one of the numeric fields to end the loop.

BUILDING BLOCK 11 — Terminating a Loop – The Sum-of-the-Fields Technique

FUNCTION • The *sum-of-the-fields* technique directs the computer to add several data fields together, test the total, and terminate a loop if the total equals zero.

APPLICATIONS • Terminating a loop by the sum-of-the-fields technique is convenient in cases where all of the fields of a record are being used for information necessary for processing. The programmer places a blank record at the end of the data set. The program is instructed to add the values in several fields together as it reads in each record. If the total is greater than or less than zero, processing continues or the loop repeats. If the total is equal to zero, it is assumed to be the end-of-file record (blank record) and the loop terminates.

This technique can also be used in many languages to terminate loops which perform mathematical processes or data manipulation routines. Some compilers, though, are designed to consider blank data fields as an error condition and may terminate execution of a program or branch to an error routine.

LOGIC • A loop being terminated by the sum-of-the-fields technique contains an arithmetic operation and a conditional branch. The arithmetic operation adds the values in the specified fields together and the conditional branch tests the total to see if it is equal to zero. If it is not, the loop continues or is repeated. If it equals zero, the program branches to another point in the program (see Figures 6.16A and 6.16B).

Since many compilers will interpret a blank card as containing zero data, care must be taken to see that the fields being added together would not normally contain all zeroes.

Placement of the arithmetic operation and conditional branch must be considered to

Computer Algorithms and Flowcharting

TERMINATING A LOOP: SUM-OF-THE-FIELDS TECHNIQUE

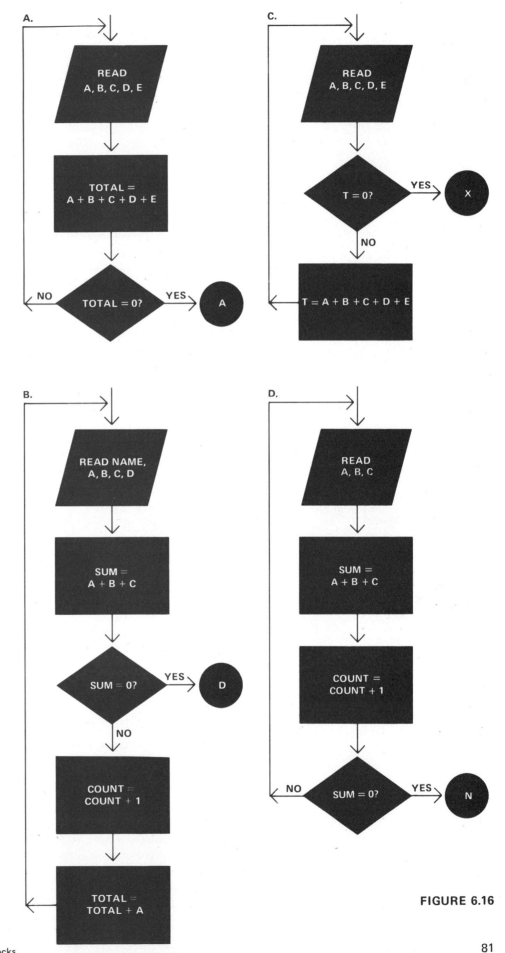

FIGURE 6.16

Chapter Six: Program Building Blocks

avoid errors. In Figure 6.16C the total is tested before it has been calculated. On some computers this would be considered an undefined variable. On others it would be equal to zero the first time through the loop and the loop would terminate.

In Figure 6.16D the incrementing statement of the counter has been placed in the wrong location. Consequently, the blank record at the end of the file will be counted as a processed record.

The arithmetic operation for adding and the IF statements are used in FORTRAN, BASIC, and COBOL to perform this technique.

Exercises

1. Briefly state the logic followed when terminating a loop with the sum-of-the-fields technique.
2. When would this technique be used?
3. What are the two major problems that may arise when this technique is used?
4. Flowchart a loop that is terminated by the sum-of-the-fields technique.
5. Flowchart a loop that reads in and counts records and that terminates with the sum-of-the-fields technique.

Terminating a Loop Using a Counter

FUNCTION • This technique directs the computer to terminate a loop when a counter located within the loop reaches a specified value.

APPLICATIONS • Using a counter to terminate a loop has several objectives and advantages. The programmer can set a limit, within the program, on the number of times a loop is executed. In Figure 6.17A, the loop is limited to 100 repetitions. Control will move to the next process after 100 names have been read in.

A counter can be used to limit the number of times an output loop is executed and, at the same time, number the records, pages, and lines as they are printed out by the loop. In Figure 6.17B, the inner loop allows 30 lines to be printed on a page, and the outer loop numbers the pages and stops after ten pages have been printed out.

A counter may be used to limit the number of times each path in a multiway branch is repeated. In Figure 6.17C, the loop for each class will terminate after ten names have been added.

Counters can be combined with trailer records to produce an algorithm such as the one shown in Figure 6.17D. A data file with an undetermined number of records is read in and counted. A test for trailer values terminates the read loop. The value of the counter is used to limit the number of times other processes and the output loop are executed.

LOGIC • Terminating a loop with a counter involves initializing a counter, incrementing it, and testing it with a conditional branch. Generally, this is what happens: the computer reaches the initializing statement first and sets the beginning value of the counter. It enters the loop, performs the indicated process, and increments the counter. A conditional branch tests the value of the counter against the test condition. The test condition may be a constant (set by a programming statement), a variable (determined elsewhere in the program), or a mathematical computation. If the conditions of the test

TERMINATING A LOOP USING A COUNTER

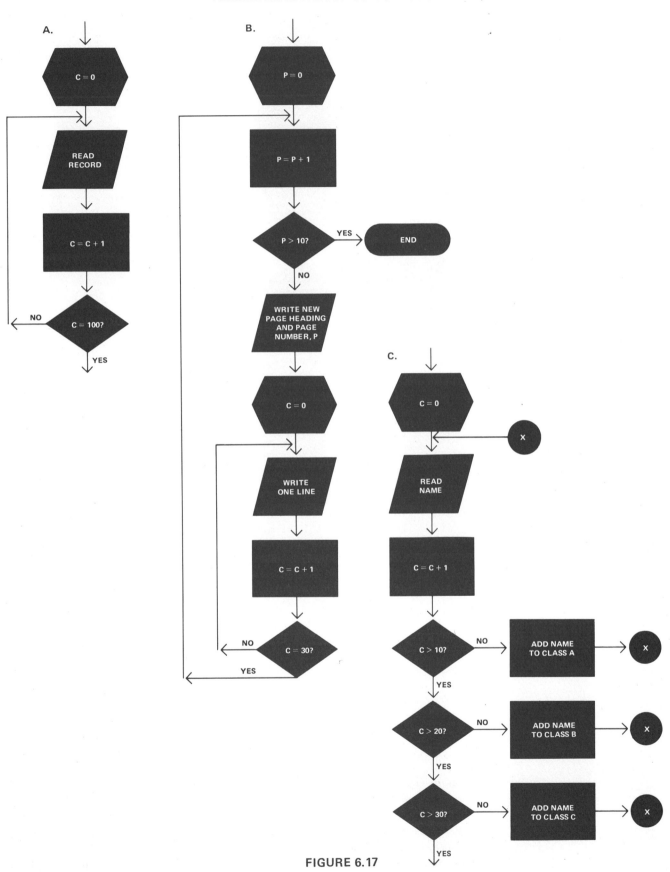

FIGURE 6.17

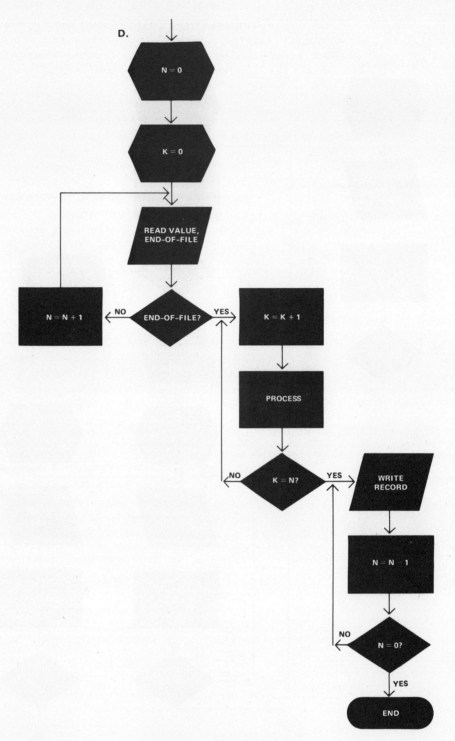

FIGURE 6.17 (CONTINUED)

are not met, the loop is repeated; the process is performed again and the counter is incremented and tested again. When the counter meets the conditions of the test, the computer terminates the loop.

Figure 6.17B shows two nested loops, each terminated by a counter. The computer initializes the page counter P before entering the outer loop. Then P is increased by one and tested to see if it is greater than ten. If it is, program execution terminates. If it is not, the new page heading sequence is performed and the value of P is printed out as the page number. Notice the placement of the conditional test. Since it will be executed before the WRITE instruction, it tests for a value greater than ten. (If it were equal to ten, the program would terminate before the tenth page were printed.) C, which counts the number of lines on a page, is initialized to zero and the computer enters the inner loop. One line is printed out. C is increased by one and tested. If it is not equal to 30, the inner loop is repeated. Another line is written and C is increased and tested again. When C is equal to 30 (30 lines have been written on one page), the loop terminates and control returns to the top of the outer loop to start the next page.

Figure 6.17C shows a program with several conditional branches. Each branch terminates a loop after ten repetitions. The counter C is initialized to zero, a name is read in, and the counter is increased by one. The computer tests the counter to see if it is greater than ten. If it is not, the name is added to the list in Class A and control returns to the top of the loop to read in another name, increase C, and test its value again. When the counter is greater than ten, control drops to the next instruction, which is another conditional branch. This branch adds names to Class B until the counter is greater than 20. The next branch adds names to Class C until the counter is greater than 30. Control then drops to the next statement and processing continues.

Figure 6.17D illustrates an algorithm that uses two counters and a trailer record to control three sequential loops. First the counters N and K are initialized to zero. Then the computer enters the first loop. A record is read in and a field named EOF (end-of-file) is tested to see if it is the trailer value. If it is not, N is increased by one and the loop is repeated. The next record is read in and tested. When the end-of-file condition is reached, the loop terminates and the program branches to the next process, which is the second sequential loop. At this point, N is equal to the number of records read in. The counter K is increased by one, the process is performed, and K is tested to see if it is equal to N. If it is not, the loop is repeated. When K is equal to N, the process has been performed the same number of times that records have been read in. The loop terminates and control moves to the next part, the third sequential loop, which prints out N lines. One line is printed out and N is decreased by one and tested. If it is not equal to zero, the loop is repeated. Another record will be printed out, and N will again be decreased and tested. When N is equal to zero, the loop and the program terminate.

Notice the placement of the incrementing statement in the last loop. Since N is decreased after the process is performed, the loop will not be executed when N equals zero. If it had been necessary to number the records as they were being printed out, the same type of counter as K could have been used instead of the negative N counter.

In some instances, to avoid possible error situations, the relational operators for greater than (>) or less than (<) should be used instead of equal to (=). This is especially important when the increment is a value other than one.

See Building Block 7 for more details on designing and programming counters.

Counters used for terminating loops are programmed in FORTRAN with the assignment and IF statements. BASIC uses the IF and LET statements. In COBOL, the VALUE clause, MOVE, IF, and arithmetic statements are used.

Exercises

1. State the logic followed in terminating a loop with a counter.
2. When is a counter used to terminate a loop? Why?

3. How would a program read in and write out an undetermined number of records within loops?
4. Flowchart a program module that reads in 100 records.
5. Flowchart a program that reads in a data file with an undetermined number of records and prints them out, using a WRITE loop terminated by a counter.
6. Flowchart a program with three sequential loops. Use the sum-of-the-fields technique to limit one loop, a positive counter to limit another, and a negative counter to limit the third.

Limited Loops – A Language Feature

FUNCTION • The limited loop is a language feature that directs the computer to repeat a specific sequence of instructions a certain number of times. It is a language statement that conveniently and automatically performs the functions a counter and a conditional branch perform to terminate a loop. This statement has a built-in counter.

APPLICATIONS • Many languages contain a special statement that provides the programmer with a convenient and flexible means of controlling a loop. The number of repetitions, the initializing and incrementing values, and the range of the loop are set by the programmer. They may be constant values or variables. Limited loops are practical to use in most situations that do not lend themselves to the use of a trailer or to the sum-of-the-fields method of controlling loops. They are sometimes used in conjunction with trailer records to control, count, and terminate loops that read in records from a file.

LOGIC • The language statement used to automatically limit loops contains a built-in index and counter. In FORTRAN the DO statement is used. It gives the number of the last statement included within the loop, the name of the index, and the initial, incrementing, and ending values. A CONTINUE statement is sometimes used to mark the end of the loop:

```
40      DO 70 I=1,10,1
50          READ (1,60) AMOUNT
60          FORMAT (F8.2)
70      CONTINUE
```

In BASIC the FOR/NEXT statements are used. The processes to be repeated are included between these two statements. The FOR statement gives the index name and the initial, incrementing, and ending values. The NEXT statement marks the end of the range of the loop:

```
30      FOR I = 1 TO 10 STEP 1
40          READ A
50      NEXT I
```

In COBOL the PERFORM statement is used. It tells the computer which instructions to repeat and how many times they are to be repeated:
PERFORM READ-AMOUNT 10 TIMES.

Figure 6.18A shows the basic steps performed by this language feature. The symbols in solid lines represent steps that are written as coded language statements by the programmer. The symbols in broken lines represent the steps performed automatically by the computer when executing the limited loop instructions.

The computer sets the initial value of the index to the specified value. Then it performs the processes included within the loop. It increases the value of the index by the increment and then tests it to see if it is greater than the ending, or limiting, value. The loop is repeated and the index is incremented until the value of the index is greater than the test condition. At this point the loop terminates and control drops to the next instruction. The value of the index may be accessed and used at other points in the program.

LIMITED LOOPS

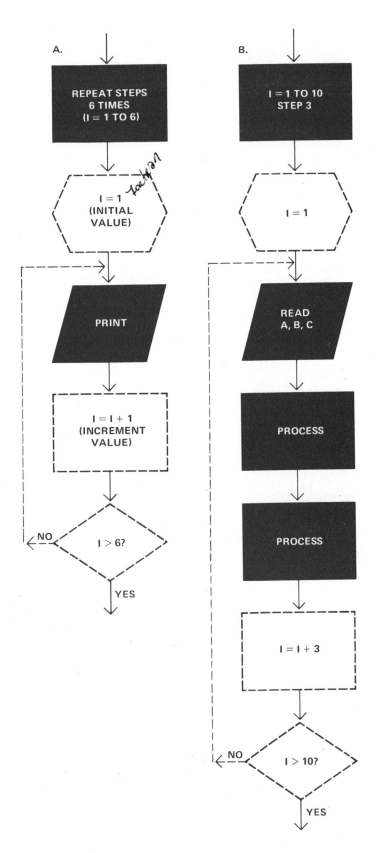

FIGURE 6.18

Chapter Six: Program Building Blocks

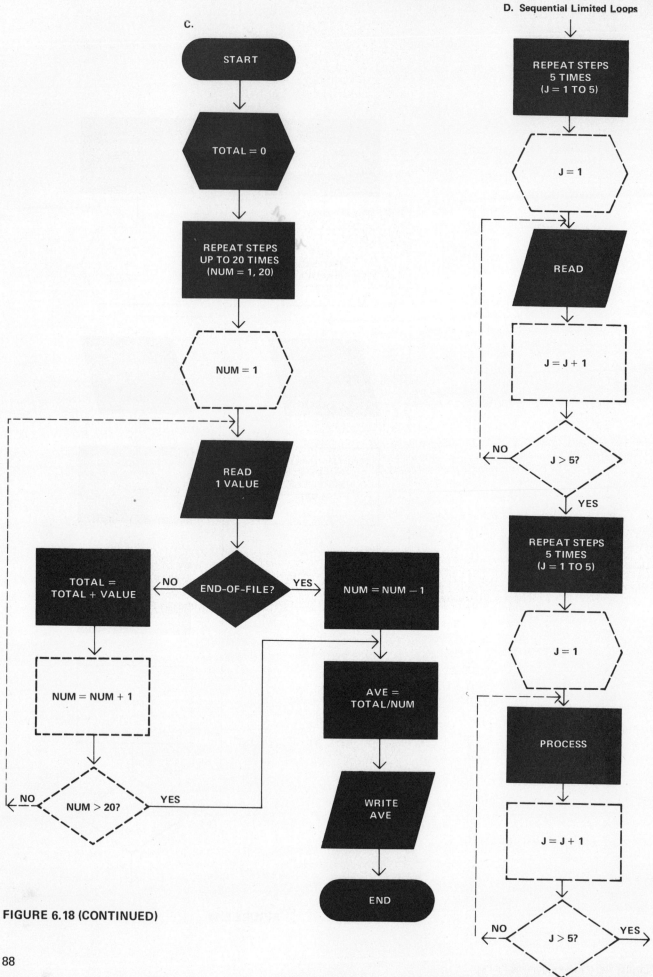

FIGURE 6.18 (CONTINUED)

A. Logic

NESTED LIMITED LOOPS

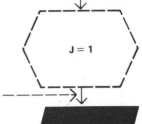

B. Output Data

I	J
1	1
1	2
1	3
1	4

END OF OUTER LOOP 1

2	1
2	2
2	3
2	4

END OF OUTER LOOP 2

3	1
3	2
3	3
3	4

END OF OUTER LOOP 3

FIGURE 6.19

Chapter Six: Program Building Blocks

The incrementing value of the index does not always have to be one. It can be any value expressed as an integer or a decimal. It tells the computer how much to change the value of the index each time the loop is executed. Figure 6.18B flowcharts a problem where the increment is three and the limiting value is ten. The first time through the loop I (the index) equals one. It is then increased by the increment value (three) and tested against the limit (ten). Since I now equals four, which is less than ten, the loop is performed a second time. I is increased to seven, still less than ten, so the loop performs a third time. I is increased to ten, but since it still is not greater than ten, the loop performs a fourth time. When I increases to 13 and satisfies the test condition, the loop terminates.

In Figure 6.18C, the final value of the index NUM represents a count of the number of values read in. It is accessed later in the program when the average is calculated. In this example, the limited loop is used in conjunction with a trailer record to allow the program to handle fewer than 20 values. The loop will terminate when the EOF record is read in or when NUM is greater than 20, whichever occurs first. NUM will reflect the number of values read in. Notice that if EOF occurs first, NUM is decreased by one, since the trailer record has been included in the tally.

Figure 6.18D shows two sequential limited loops. The first loop is executed five times and then the second loop is executed five times. Since the loops are independent of one another, the same name was chosen for both indexes; different names could be used.

Figure 6.19A is a flowchart of a program with two nested loops. Each loop controls a statement that prints out the current value of the indexes. Since the loops are interdependent, each must have its own unique index name.

The first time through the outer loop, I equals one, and J equals one the first time through the inner loop. The PRINT statement outputs these values. J is increased to two and tested against the limit. The inner loop is repeated and the PRINT statement outputs the current value of I (1) and J (2). J is increased and tested and the loop repeats. I is still equal to one and J is now equal to three. The fourth time through the loop, the PRINT statement outputs *I=1* and *J=4*. J is increased to five, which satisfies the test condition of the conditional branch, and the loop terminates. Control returns to the outer loop to print out a note that the first pass of the outer loop is now completed. I increases to two and is tested, and control returns to the beginning of the outer loop. The inner loop will output four more lines with I equal to two, and the outer loop will note the end of its second repetition. The cycle will be repeated a third time, with I equal to three. Then I increases to four, exceeding the limit, and the loop terminates. Figure 6.19B shows the output from Figure 6.19A.

Exercises

1. Briefly describe the logic followed in limited loops.
2. What are the advantages of limited loops?
3. Flowchart a limited loop that reads in 50 records. Show all steps in the loop. Redraw the flowchart to show only the steps that need to be coded.
4. Flowchart a program that reads in a data file with an undetermined number of records and prints them out with a limited loop.
5. Flowchart a program with nested loops. Include print statements in each loop. Have the inner loop execute twice and the outer loop five times. Show the output.
6. Flowchart a program with nested sequential loops. Have the outer loop execute twice, an inner loop three times, and a second inner loop four times. Include print statements and show the output.
7. Flowchart a program with three nested limited loops. Have the inner loop execute three times, the middle loop four times, and the outer loop twice. Include print statements and show the output.

BUILDING BLOCK 14

One-Dimensional Arrays

FUNCTION ● It is frequently necessary in programming that all the items or records in a data file be available in storage at the same time. A program that reads in a large number of items and stores each under a different name would be quite cumbersome and inefficient. Most languages offer facilities for directing the computer to automatically read in a list of data items, assign each a unique name, and store them in sequential locations in memory. This storage arrangement is called an *array*.

APPLICATIONS ● Processing long lists of data items is done in one of two ways. A program can read in one item at a time, store it under a variable name, and perform one or more operations upon it. Subsequent items are read in, stored under the same variable name (the new values replace the old) and processed. This method requires only a minimum amount of storage space for processing an entire data file—enough to hold one record.

Other algorithms require that the entire file be directly accessible in storage at one time. This would be true of sorting, alphabetizing, or searching routines, as well as of many mathematical procedures and some listing operations.

For example, a program that prepares a telephone directory utilizes an array. Records containing names, addresses, and phone numbers are read into the computer in no specific order and stored as an array. The program re-sorts the data into alphabetic order and prints out the rearranged array. (The data can also be sorted by geographic locations or area code.)

A program that reads in and adds several test grades for each student in a class and prints out the totals in numeric order also utilizes an array.

Two or more parallel arrays are often used in data processing. For instance, names can be stored in one array and old balances in corresponding positions in a second one.

LOGIC ● ANSI defines an array as an arrangement of elements in one or more dimensions. It may be thought of as a row or column of related sequential storage locations. One name is assigned to the group as a whole, and each location within the group has its own number. Each location, therefore, has a unique name, different from the other locations in the array. Any location can be referenced directly in a program by using its array name and location number. (This is similar to houses on a street having unique addresses but still sharing the name of the street.)

The location number of each position in the array is called the *index* or *subscript*. ANSI defines an index as a symbol or numeral used to identify a particular quantity in an array of similar quantities. The subscript or index is shown in parentheses after the array name. STOCK(1), ITEM(10), and BAL(3) indicate specific locations within an array.

Most languages have facilities for reading data items into an array automatically. The program must assign a name to the array, reserve the required number of storage spaces, and provide a means of changing, or incrementing, the value of the index or subscript.

In FORTRAN, DIMENSION statements are used to reserve storage spaces for arrays. BASIC uses DIM statements and COBOL uses OCCURS clauses.

Limited loops or loops with counters are used for this purpose. The variable name assigned to its index is also used as the name of the array subscript. The computer will substitute the current value of the index for the value of the array subscript each time the statement is executed.

This is how it is done with a limited loop in FORTRAN:

```
      DIMENSION NUMBER (10)
      . . .
      DO 50 I=1,10
      READ (1,40) NUMBER (I)
   40 FORMAT (I3)
   50 CONTINUE
```

LOADING ARRAYS

Loading an Array With a Limited Loop

A.

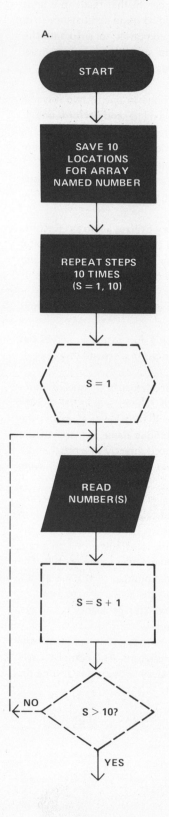

FIGURE 6.20

A BASIC program with a limited loop:

```
10    DIM N (10)
...
50    FOR I=1 TO 10
60        READ N (I)
70    NEXT I
```

Using a counter in BASIC:

```
10    DIM N (10)
...
50    LET I=1
60    READ N (I)
70    LET I=I+1
80    IF I > 10 THEN 100
90    GO TO 60
```

Loops and counters are used in COBOL programs to load and manipulate arrays. Figure 6.20A shows how ten numbers are loaded into an array by a limited loop. First the program directs the computer to reserve ten consecutive locations under the name NUMBER. Then control enters the limited loop. The indexes of both the loop and the array are named S. The initial value of S is set to one and the ending value to ten. The first value (456) is read in, and since S equals one, it is stored in the first location of the area reserved for NUMBER. S increases to two and is tested, and the control returns to the READ statement. The next number (128) is read in, and since S equals two, it is stored in the second position in NUMBER. S increases to three and the next value is stored in position three of NUMBER. This continues until all ten values have been read in and stored. Then S increases to 11, greater than the ending limit, and the loop terminates. Figure 6.20B shows the ten values to be read in and Figure 6.20C shows them in storage under the name NUMBER.

Figure 6.20D shows the same list of ten values being read into the array with a counter. (A test for trailer record terminates the READ loop.) The next part of the flowchart shows how each item in the array is processed sequentially by changing the value of the index. In this example, a counter controls this loop also. The index I is initialized to one. The value in location one of the array NUMBER is printed out. The index increases to two and is tested against the final value of S (ten). If I is less than or equal to S, the loop repeats. The value in location two of NUMBER is output. I increases to three and is tested. After the value in location 10 has been printed, I will increase to 11 and the loop will terminate.

Figure 6.20E is a flowchart of a program that uses a limited loop to read five old balances into an array named OLD. The index is named N and has a starting value of one and a limit of five. As each old balance is read in, it is also added to a running total (OLDTOT). It is referenced in the arithmetic statement by its unique name—the array name and its subscript. The next portion of the flowchart calculates the new balance for each element in the array. A limited loop is again used to change the value of the index and drive the computer one item at a time through the array. The last portion of the flowchart prepares the output. The elements are printed out starting at the bottom of the array and moving to the top. This is accomplished by initializing the index to five and instructing it to decrease by one. Figure 6.20F shows the input and output for this program.

Figure 6.20G shows how two parallel arrays are loaded and manipulated with the same limited loop. The subscripts for both arrays are controlled by the index I. Five storage spaces are first reserved for each array. Items are read in two at a time (there are two items

Computer Algorithms and Flowcharting

Loading an Array With a Limited Loop (Continued)

Loading and Manipulating Arrays with Counters

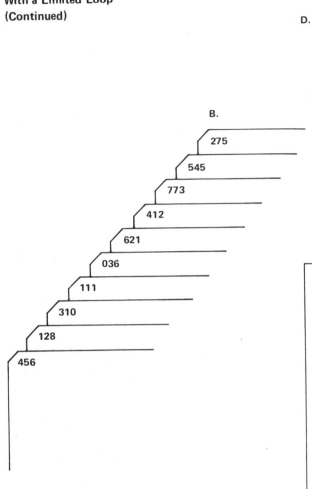

C. Array — Number

Position Stored In	Data Item	Unique Name
S=1	456	NUMBER(1)
S=2	128	NUMBER(2)
S=3	310	NUMBER(3)
S=4	111	NUMBER(4)
S=5	036	NUMBER(5)
S=6	621	NUMBER(6)
S=7	412	NUMBER(7)
S=8	773	NUMBER(8)
S=9	545	NUMBER(9)
S=10	275	NUMBER(10)

FIGURE 6.20 (CONTINUED)

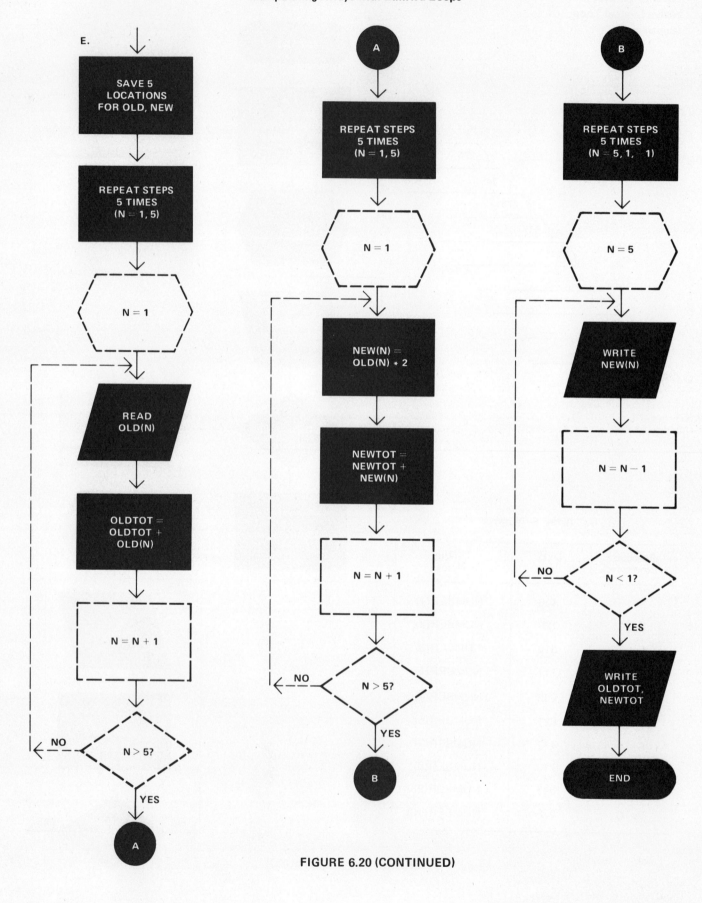

FIGURE 6.20 (CONTINUED)

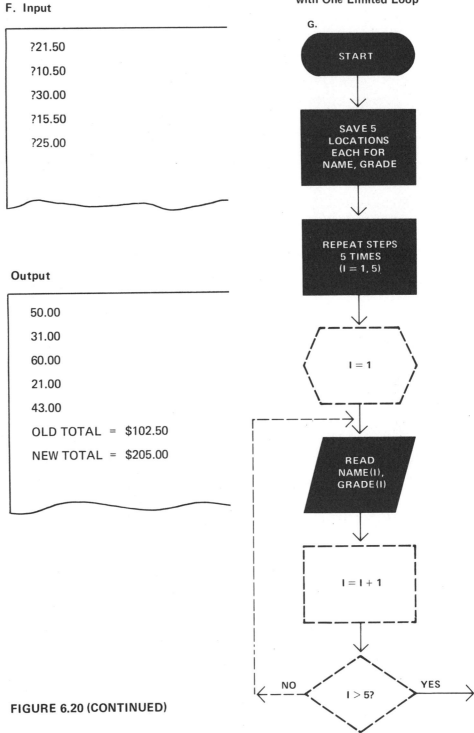

FIGURE 6.20 (CONTINUED)

on each record). The first item is placed in the array NAME, and the second into the array GRADE. Since I equals one, both items go into the first locations in the arrays. I increases to two, and the next record, containing two items, is read in. The name is placed in the second location of NAME and the grade in the second location of GRADE. I increases to three and the next two items are loaded into corresponding locations in the two arrays.

Any item in either array may be accessed later in the program independently of the others by its unique name. The arrays may later be manipulated singly, or together. Figure 6.20H shows the input for this program and how the values appear in storage.

H. Input

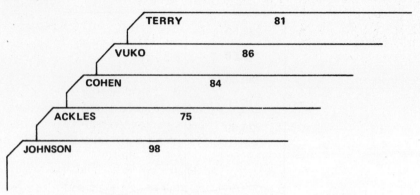

In Storage:

Array — Name   Array — Grade

Position	Data Item	Unique Name	Data Item	Unique Name
I=1	JOHNSON	NAME(1)	98	GRADE(1)
I=2	ACKLES	NAME(2)	75	GRADE(2)
I=3	COHEN	NAME(3)	84	GRADE(3)
I=4	VUKO	NAME(4)	86	GRADE(4)
I=5	TERRY	NAME(5)	81	GRADE(5)

FIGURE 6.20 (CONTINUED)

Exercises

1. Define an array.
2. When is an array used in programming?
3. How are arrays reserved, loaded, and referenced?
4. Explain the logic followed in loading an array with a limited loop.
5. Flowchart a program that loads ten names into an array with a limited loop.
6. Redraw the program in Exercise 5 to load the array with a counter.
7. Flowchart a program that reads 15 numbers into an array, adds five to each number, and prints out the new array, using three limited loops.
8. Redraw the program in Exercise 6 to use only two limited loops.
9. Flowchart a program that reads in and prints out the following data:

NAME	AGE
Nancy	8
Lea	7
Carmen	9
Richard	10
Thomas	8
Steven	6
Larry	9
Robin	9

Two-Dimensional Arrays

FUNCTION ● A computer can be directed to automatically store lists of data in tabular or matrix form. Storage locations are arranged in rows and columns. This arrangement of horizontal and vertical dimensions is called a two-dimensional array.

APPLICATIONS ● Two-dimensional arrays are used whenever it is necessary or convenient to store data in the computer in the form of tables or parallel columns. Two-dimensional arrays are used in algorithms that perform table lookups, searches, and sorts.

For example, a program that calculates discounts on merchandise purchased can use a table which lists all the prices for the relevant conditions. The program reads in and stores the two-dimensional table. Then the information on an account is read in and processed by the program. At some point the program accesses the data in the table, locates the pertinent information, and finishes processing the account. Then the program reads in each of the next accounts and repeats the process, using the same table in storage.

Many programs that perform mathematical operations require the data to be in two-dimensional form. In other programs, tables contain information necessary for the calculations being performed. Two-dimensional arrays are used in programs that perform matrix manipulations.

Alphabetic data is usually stored in one- or two-dimensional arrays for FORTRAN programs. All kinds of text, lists of names, and other alphanumeric information are easily manipulated and processed in this manner.

LOGIC ● ANSI defines an *array* as an arrangement of elements in one or more dimensions. A two-dimensional array may be thought of as a group of related storage locations arranged in parallel columns (or parallel rows). One name is assigned to the group as a whole, and each location within the group has its own number. Each location, therefore, has a unique name. Any location can be referenced directly in a program by using its array name and location numbers.

The location numbers, which signify the position of an element in the array, are indicated by two subscripts (or indexes), one representing the column of the element, and the other, the element's row. The subscripts are shown in parentheses after the array name. TABLE(1,4) and PRICE(3,2) refer to specific locations within an array.

Many languages have facilities for automatically reading data items into two-dimensional arrays. The program must assign a name to the array, reserve the required number of rows and columns for storage, and provide a means of changing and incrementing the values of the indexes or subscripts.

In FORTRAN, DIMENSION statements are used to reserve storage spaces for arrays. In BASIC, DIM statements are used, and in COBOL, OCCURS clauses perform this function. Nested limited loops or nested loops with counters are used to increment the indexes. The variable name of each index is the same as one of the subscripts. Thus, one loop controls the rows and another controls the columns. Each time the statement is executed, the computer substitutes the current value of each index for the value of the array subscript that bears the same name.

Figure 6.21A illustrates the loading of a two-dimensional array. First the program saves 12 storage locations arranged in three rows and four columns. Control enters the outer loop and initializes I to one. The index I indicates which row is being loaded. Then control enters the inner loop and initializes the index J to one. J indicates which column is being loaded. The first data item (11) is read in and stored in location 1,1, where column 1 and row 1 intersect. J increases to two, and the second item is loaded into the location at the junction of row 1, column 2. The third item goes into row 1, column 3, and the fourth into row 1, column 4. At this point, J increases to five, is greater than the limit, and the first execution of the inner loop terminates. Control returns to the outer loop. I increases to two and is tested against the limit (three). Since it is less than three, control returns to the beginning of the outer loop. The inner loop will be executed a second time, with I equal

Chapter Six: Program Building Blocks

Loads Horizontally

TWO-DIMENSIONAL ARRAYS

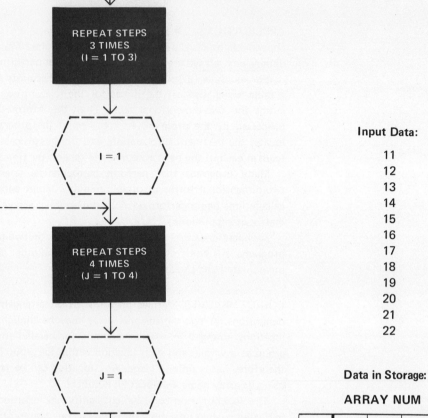

FIGURE 6.21

Computer Algorithms and Flowcharting

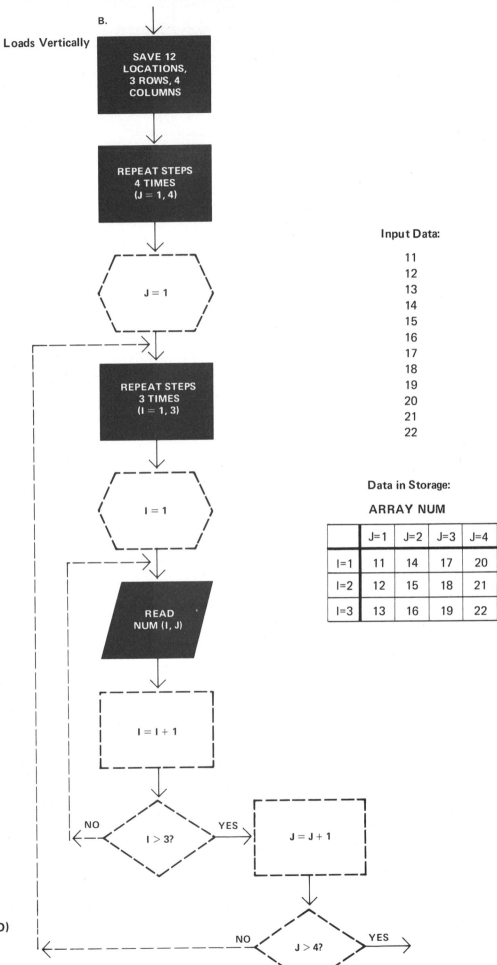

FIGURE 6.21 (CONTINUED)

Chapter Six: Program Building Blocks

Processing Two-Dimensional Arrays

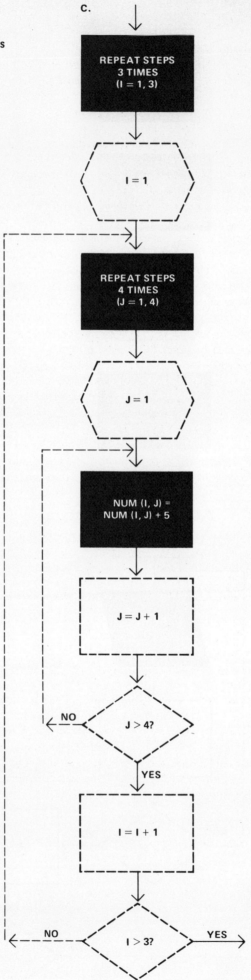

Data in Storage After Processing

ARRAY NUM

	J=1	J=2	J=3	J=4
I=1	16	17	18	19
I=2	20	21	22	23
I=3	24	25	26	27

FIGURE 6.21 (CONTINUED)

to two. J resets to one and loads the next data item into location 2,1, where row 2 intersects with column 1. The next items go into locations 2,2; 2,3; and 2,4. J increases to five and the inner loop terminates the second execution. I increases to three, less than the limit, and the entire cycle is repeated again—the last four items are loaded into 3,1; 3,2; 3,3; and 3,4.

In Figure 6.21B the same array loads vertically. The subscript J, which controls the columns, is placed in the outer loop. The subscript I, which controls the rows, is placed in the inner loop. Since the inner loop is executed three times for each execution of the outer loop, the items will load vertically. Then J will increase to two and the inner loop will be repeated three more times, loading the second column. J increases again and the third column is loaded, followed by the fourth. At that point the outer loop terminates.

Figure 6.21C illustrates how items in two-dimensional arrays are processed sequentially. In this program, five is being added to each element in the array. The program enters the outer loop and initializes I to one. The inner loop is entered and J is set to one. Then the data item in location NUM(1,1) is replaced with a new value, equal to the old number plus five (11+5=16). J increases to two, is tested, and the inner loop repeats. The item in location NUM(1,2) is increased by five (12+5=17) and replaced. J increases to three and the item in location NUM(1,3) is processed. When J is greater than four, the inner loop terminates. I increases to two and the next four items are processed. This continues until the conditions for ending the outer loop are met and the procedure terminates.

Any individual item could, of course, be accessed by the program and processed by itself by using its array name and subscripts. For example, the value 21 in Figure 6.21C could be used later in the program by referencing NUM(2,2). NUM(1,4) would access the value 19, and NUM(3,2) the value 25.

In some languages only numeric data can be stored in two-dimensional arrays. Others permit alphanumeric or numeric data to be manipulated in two-dimensional arrays. Most languages require that an array be defined as either alphanumeric or numeric.

Exercises

1. Define a two-dimensional array.
2. How are two-dimensional arrays loaded? How are the items referenced?
3. Explain the logic followed in loading two-dimensional arrays with nested limited loops.
4. Flowchart a program that loads a two-dimensional array with five rows and six columns. Show the data as it would look in storage.
5. Flowchart a program that loads the array in Exercise 4 vertically.
6. Flowchart a program that loads 12 numbers into a two-dimensional array, adds 10 to each number, and then prints out the numbers in a vertical list.
7. Flowchart the program in Exercise 6 using both an array with different dimensions and a different means of outputting the list.

Chapter SEVEN
Applied Programming Logic

This chapter presents a group of programming problems representative of those found in data processing. The programming techniques, algorithms, and program logic commonly used to solve these kinds of problems are illustrated and explained in terms of the building blocks discussed in Chapter Six.

Each unit states a problem similar to the kinds of problems a programmer might actually encounter. Specifics of the problem are analyzed and outlined, and the program is categorized according to purpose and type. A flowchart presents a graphic illustration of the algorithm selected for solution. A written narrative traces the path the computer will follow as it executes the programming steps. Minute programming details, especially those relevant only to specific languages, are avoided to prevent the major steps in the logic from being obscured.

The reasoning behind the selection of one technique or algorithm over another is discussed whenever pertinent. In many instances, several algorithms would be equally successful in solving a problem — indeed, when problems of some complexity are involved, it is rare to find two programmers who flowchart and code the problem exactly alike.

The examples range from those employing relatively simple algorithms, involving only a few building blocks, to more complex problems that illustrate the range of the computer's capabilities.

UNIT 1 SIMPLE CALCULATION AND REPORT PREPARATION PROBLEM

PROBLEM • The Plush Carpet Company wants a computer program that will calculate the retail selling price of carpeting, given the room measurements and cost of materials. Input will be a set of punched cards, each containing the account name, room size in square feet, and the cost of the carpeting selected.

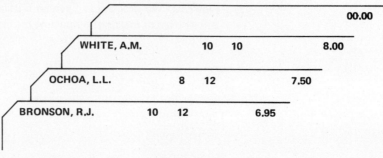

FIGURE 7.1

Output should be a one-page report with appropriate headings:

ACCOUNT NAME	SQUARE YARDAGE	SELLING PRICE
BRONSON, R. J.	13.3	$92.44
OCHOA, L. L.	10.67	$80.03
WHITE, A. M.	11.1	$88.80

FIGURE 7.2

The program should be able to handle a variable number of input records.

SOLUTION ● This problem illustrates the processing of a data file one record at a time. A simple calculation is performed on each record and a one-page report is prepared. Since the input, arithmetic, and output processes are to be repeated on each set of data, they are included within a loop. Using a trailer-record test to terminate the loop allows a different number of records to be processed each time the program is executed.

The first procedure the program will perform is to write the column headings for the report. The next step reads in the first data record containing the account name, room dimensions, and carpet cost. In this example, the cost field is tested to see if it equals 00.00 — the trailer record. If it does not, control continues to the next step.

An arithmetic process calculates the square yardage needed for the job. This is found by figuring the area of the room in square feet (DEPTH * WIDTH) and dividing the product by nine. The next arithmetic process calculates the selling price by multiplying the yardage by the cost of a square yard of carpet.

The last process writes the name of the account, the total yardage calculated, and the selling price under appropriate column heads. An unconditional branch directs control back to the beginning of the loop to read in the next record. Yardage and selling price are calculated for the next account, and the data is listed beneath the figures for the first account. The process continues until the trailer record containing 00.00 in the cost field is read in; at that point, execution terminates.

BUILDING BLOCKS USED:
No. 2 — Unconditional branch
No. 3 — Two-way conditional branch
No. 6 — Simple loop
No. 10 — Terminating a loop with a trailer record

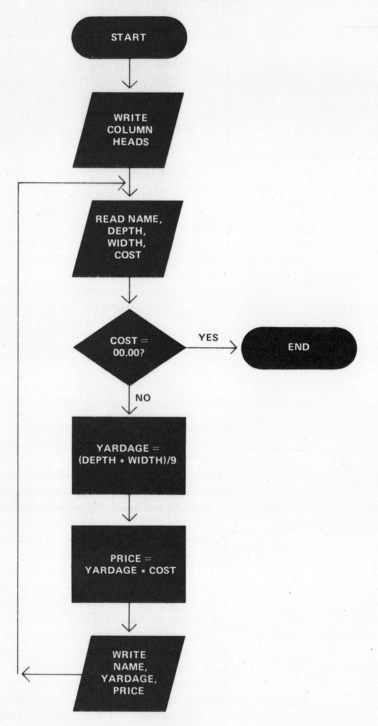

FIGURE 7.3

Exercises

1. Which symbol represents the unconditional branch on the flowchart for this problem? How else could it have been shown?
2. What would have happened if the simple loop had included the WRITE column symbol? If it had not included the READ symbol?
3. Redraw the input file and flowchart to terminate the program with a 999 trailer value in an unused field.
4. Redraw the input file and flowchart to terminate the program with the sum-of-the-fields technique.
5. Expand the flowchart so that it calculates the total price and amount of yardage estimated and prints out these totals before terminating execution.

UNIT 2 PROCESSING DATA FILES ONE RECORD AT A TIME

PROBLEM • The Irving Bottling Company needs a program that can calculate the amount due on cases of bottles returned for credit. Irving grants full credit when five or more full cases are returned, counting cases filled halfway or more as full; otherwise half credit is granted. The records from the shipping room are punched into data cards and show the account name and number of full and fractional cases. The input file looks like this:

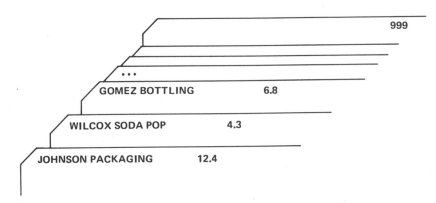

FIGURE 7.4

An output report, showing the amount of credit due, is to be generated. It is to look like this:

	RETURN REPORT		
FIRM	CASES RETURNED	FULL CREDIT GRANTED	HALF CREDIT GRANTED
JOHNSON PACKAGING	12.4	12	
WILCOX SODA POP	4.3		2
GOMEZ BOTTLING	6.8	7	
. . . .			

FIGURE 7.5

IRVING BOTTLING COMPANY

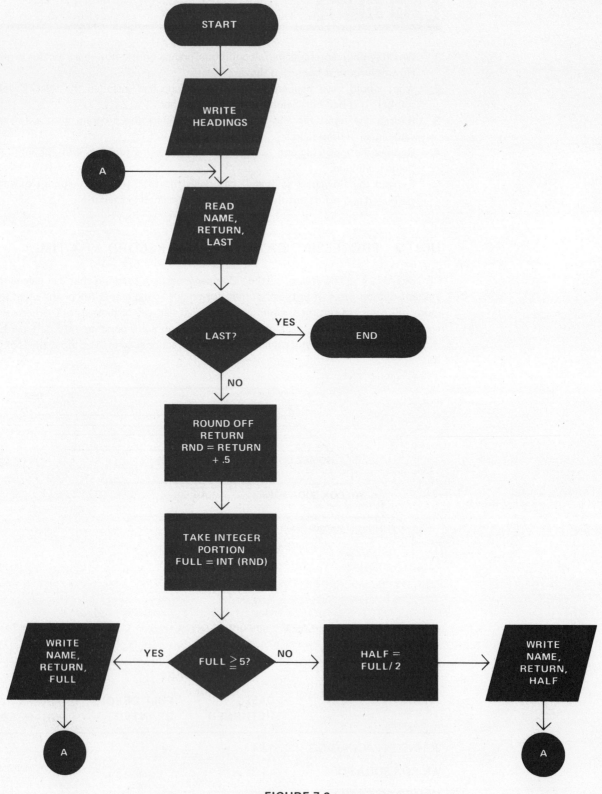

FIGURE 7.6

SOLUTION • This program illustrates performing a series of operations on a data file, one record at a time. A calculation is performed on a data field on each record, the results are tested, and one of two different output sequences is selected. A report, giving the results of the data records processed, is generated.

A simple loop and an unconditional branch are used to repeat the processes on each record in the file. A test for trailer record (999 in a reserved field) terminates the loop and execution of the program.

The program begins by printing out the headings needed for the output report. Then control enters the loop, and the first record, containing an account name and number of cases returned, is read in. The field reserved for LAST is tested to see if it contains the trailer value (999). If it does not, it is assumed to be a data record and the amount of credit is calculated for that record.

The creditable number of cases returned is found by adding .5 to the real number returned and then dropping the fractional portion of the number. If the original fraction is .5 or greater, adding .5 to it makes it greater than 1.0, raising the integer portion of the number by one. When the fractional portion is dropped, the number has been effectively rounded up. If the original fraction is less than .5, adding .5 does not increase the integer portion. When the fraction is dropped, the integer portion remains the same, rounded down. (If $x = 1.7$, $x + .5 = 2.2$ and creditable number of cases = 2.0. If $x = 4.1$, $x + .5 = 4.6$ and creditable number of cases = 4.0.)

The operation of retaining only the integer portion (or dropping the fractional part) of a number utilizes a stored function (sometimes called a subprogram) available on most compilers. In BASIC this is done with the following statement:

    60 LET F=INT(R)

In FORTRAN, a decimal value can be changed to an integer with the INT function or with an assignment statement.

The rounded total is then tested by a conditional branch to see if it is greater than or equal to five cases. If it is, the program branches to an output sequence that writes the account name and rounded total in the column for "Full Credit Granted." An unconditional branch returns control to the beginning of the loop for the next execution.

If the rounded total is less than five, the program branches to an output sequence that writes the name and the total in the "Half Credit Granted" column in the report. An unconditional branch returns control to the beginning of the loop for the next execution.

Records are processed in this manner until the trailer record containing 999 is read in and execution terminates.

BUILDING BLOCKS USED:
No. 2 — Unconditional branch
No. 3 — Two-way conditional branch
No. 6 — Simple loop
No. 10 — Terminating a loop with a trailer record

Exercises

1. How does this program differ from the program in Unit 1?
2. Why are two different WRITE sequences necessary in this problem?
3. Expand the flowchart to count the number of data records processed.
4. Expand the flowchart to calculate and print out the total number of full and half cases returned.
5. Modify the algorithm used so that only one WRITE sequence is required. (Suggestion: Have each branch assign a value of zero to the column it does not use.)

UNIT 3 PROCESSING DATA FILES BY GROUPS OF RECORDS

PROBLEM ● The sales manager of Dynamo Sales Company wants a computer program that will perform the calculations for the company's bonus incentive sales program. In the plan, the sales representatives earn points according to the type of merchandise sold. Regular merchandise earns one point per sale and special "push" merchandise earns five points per sale. Returned merchandise costs the representatives three points. The sales manager needs a daily report that will show the points earned by each representative and the company totals for each category of merchandise. The program must be able to handle an undetermined number of sales and sales representatives.

The input data for the program is on punched cards and is arranged in this order:

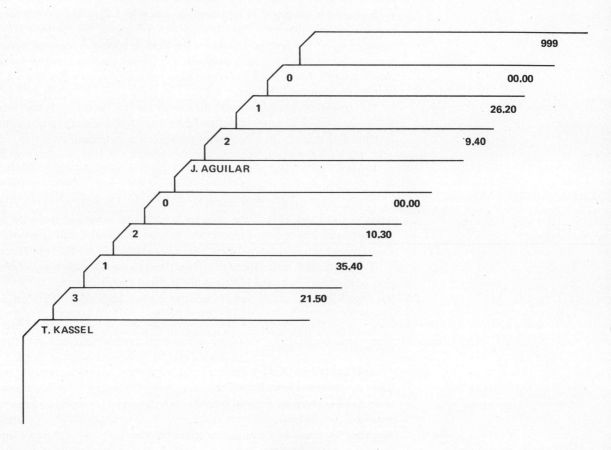

FIGURE 7.7

The name of the representative is on the first card, followed by a card for each sale that was made that day. The code for the sale category is punched into column 1 and the amount of the sale begins in column 21. The sales manager uses the following code to indicate the type of transaction:

Code 1 = Sales of regular merchandise
Code 2 = Sales of push merchandise
Code 3 = Returned merchandise

A trailer card containing zeroes in both fields is placed at the end of each representative's file. A 999 trailer record marks the end of the input file.

The output of this program should be as follows:

```
J. AGUILAR  – 3 BONUS POINTS
T. KASSEL   – 6 BONUS POINTS

TOTAL TRANSACTIONS:
    REGULAR MERCHANDISE    $61.60
    SPECIAL MERCHANDISE     19.70
    TOTAL SALES            $81.30
    LESS RETURNS            21.50
    NET SALES              $59.80
```

FIGURE 7.8

SOLUTION ● This problem illustrates the need for processing a data file by groups of records. Two mathematical operations are performed on each record in the group as it is read in. Then part of the accumulated data is output and the next group is processed. When the entire file has been processed, the remainder of the collected data is output. Parallel processing of different categories of data is also shown in this example.

The program uses an algorithm with nested loops. The outer loop controls the processing of each group of records. The inner loop processes each record in the group. A different means is used to terminate each loop. The outer loop uses a 999 trailer record and the inner loop uses the sum-of-the-fields technique. (See Figure 7.9 for the flowchart that illustrates this solution.

The program begins by initializing the mathematical fields to zero. Then it enters the outer loop and reads the first record and tests it with a two-way conditional branch to see if it is the end-of-file trailer record. If it is not, the inner loop is entered. The next record is read in and the code and amount fields summed. Another two-way branch tests the sum. If it is not equal to zero, the code field is tested to see if it is equal to one, two, or three. This can be done in several ways. A three-way branch could test CODE to see if it is less than, equal to, or greater than two. Two or more two-way branches in sequence could test for a one, two, or three value, and even test for an error value in the field.

Based on the result of the conditional tests, the program branches to the appropriate sequence of instructions. If the code is one, it indicates a sale of regular merchandise and one point is added to the representative's total points. Then the amount of the sale is added to the running total for that category. A code of two would indicate a sale of push merchandise. Five points would be added to the representative's total and the amount would be added to the accumulated total for that category. If the code is three, the computer subtracts three points from the representative's total and adds the amount of the transaction to the running total for returned merchandise. Any other value in the CODE field branches the program to an error routine.

At the end of each of these alternate paths, an unconditional branch returns control to the beginning of the inner loop and the next record is read in and processed. When the record containing all zeroes is reached, the program branches to a sequence that writes the representative's name and total points. At this point, the first group of records has been processed. An unconditional branch sends control back to the beginning of the outer loop. POINTS is initialized to zero and the program begins processing the next representative's group of records. When the trailer record (999) is reached, the program has processed all the groups of records in the file, and it branches to an output sequence that writes the accumulated totals for the three categories of transactions. Then the program execution terminates.

DYNAMO SALES COMPANY

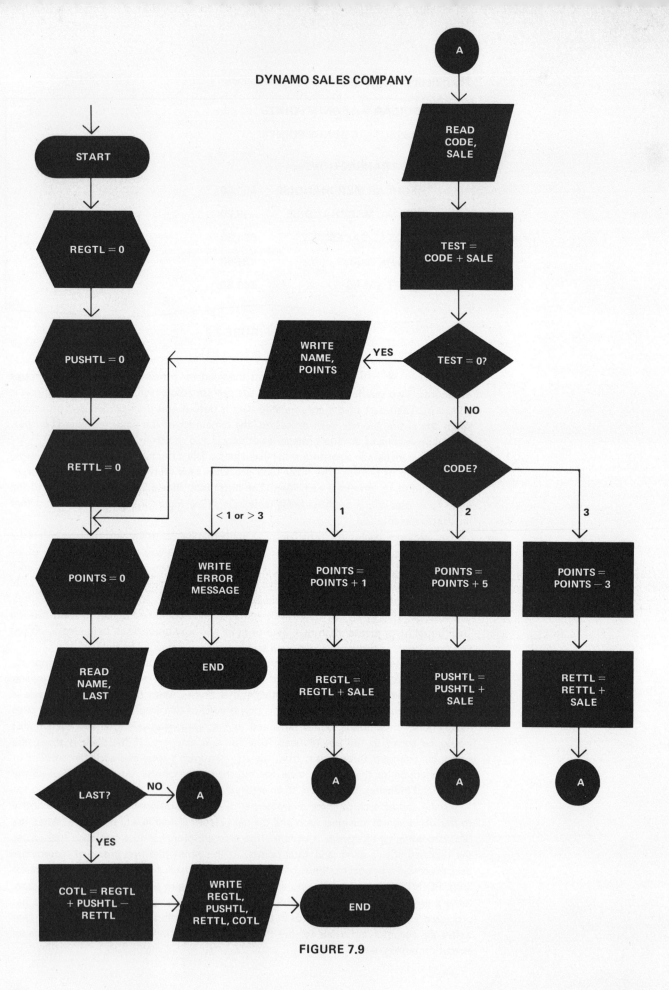

FIGURE 7.9

BUILDING BLOCKS USED:

No. 2 — Unconditional branch
No. 3 — Two-way conditional branch
No. 4 — Three-way conditional branch
No. 9 — Nested loops
No. 10 — Terminating a loop with a trailer record
No. 11 — Terminating a loop — sum-of-the-fields technique

Exercises

1. Why does the outer loop initialize only POINTS? What would happen if REGTL, PUSHTL, and RETTL were reinitialized?
2. Suppose a WRITE sequence were placed at the ends of the branches that process Code 1, Code 2, and Code 3. Would they write the category totals for each sales representative? Explain.
3. Redraw the flowchart to use three two-way conditional branches to test the code and an error condition.
4. Expand the flowchart so that the program counts and prints out the number of sales representatives listed.
5. Redraw the flowchart to include a fourth class of merchandise, month-end clearance, worth three points per sale.

UNIT 4 CLASSIFYING DATA WITH THREE-WAY CONDITIONAL BRANCHES

PROBLEM ● The Sphinx Manufacturing Company produces a large number of parts each year in its two plants. A percentage of these are returned to the main sales office because of defects in manufacture. Sphinx needs a report for management that will indicate the total amount of goods returned to each department. One of the two manufacturing plants has two divisions; the other has three divisions. Each division has three departments. The main office records all information relevant to the returned goods on punched cards. This file will be used by the inventory, quality control, and personnel offices.

The input file looks like this:

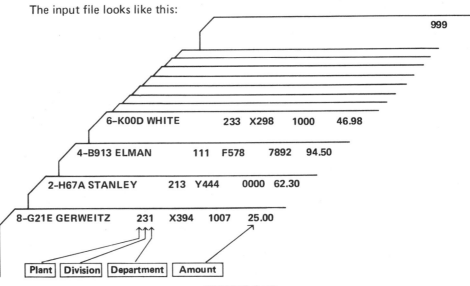

FIGURE 7.10

Chapter Seven: Applied Programming Logic

Sphinx wants the program to prepare the following report:

CHARGEBACK REPORT BY DEPARTMENT

	TOTAL AMOUNT CHARGED BACK
PLANT 1	
DIVISION 1:	
DEPARTMENT 1	128.92
DEPARTMENT 2	74.66
DEPARTMENT 3	141.00
DIVISION 2:	
DEPARTMENT 1	101.99
DEPARTMENT 2	96.58
DEPARTMENT 3	48.29
PLANT 2	
DIVISION 1:	
. . . .	

FIGURE 7.11

SOLUTION • This program illustrates classifying data with the use of multiway conditional branches. The records in a file are processed one at a time. Any data punched into the records not pertinent to this program are ignored.

A series of conditional branches is used to locate the appropriate department for the data item being processed. One branch selects the plant, a second selects the division, and the last routes control to the correct department. The names assigned to the running totals reflect the plant, division, and department numbers.

The program begins by initializing all the totals (one for each department) to zero. Then the first record is read in. The field reserved for LAST is tested to see if it contains the trailer value. If it does not, the field for PLANT is tested to see if it is equal to one or two. If it equals one, control branches to a sequence of instructions that processes all the transactions for that plant. The DIVISION field is tested and the program branches to either DIV1 or DIV2. Then the DEPARTMENT field is tested and the transaction routed to the appropriate department total. AMOUNT is added to that running total, and an unconditional branch sends the program back to the beginning of the loop to read in the next record.

A similar process is followed for the transactions related to Plant 2. When the end of the file is reached (LAST), the program writes the report for management, listing all the totals by department.

Invalid codes branch the program to an error routine.

BUILDING BLOCKS USED:
No. 2 — Unconditional branch
No. 3 — Two-way conditional branch
No. 4 — Three-way conditional branch
No. 5 — Multiway conditional branch
No. 6 — Simple loop
No. 10 — Terminating a loop with a trailer record

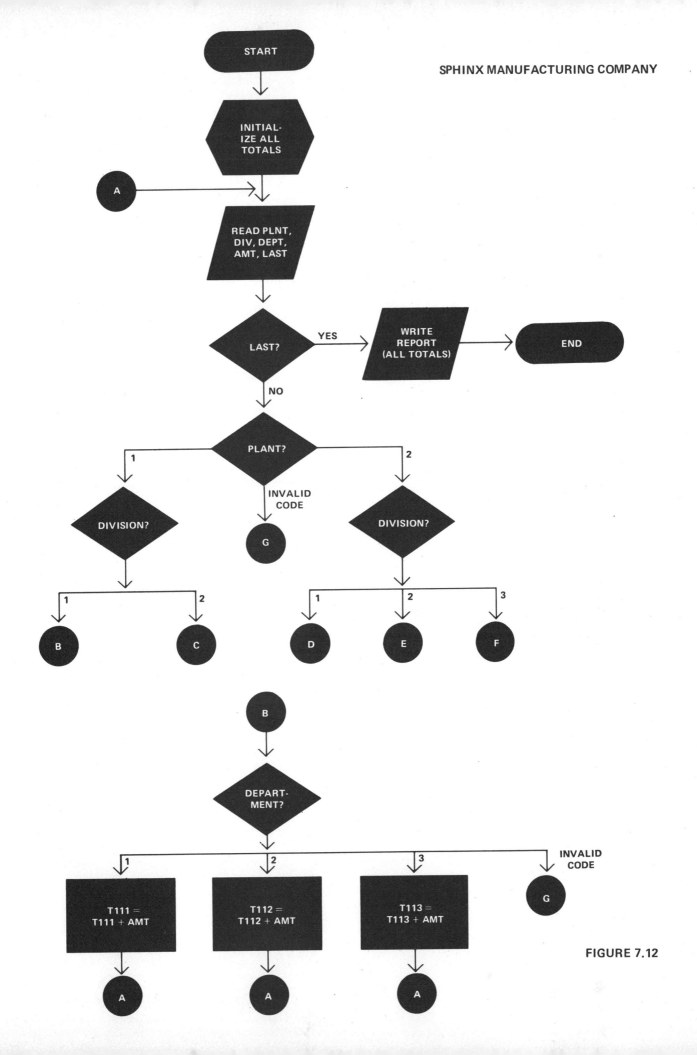

FIGURE 7.12

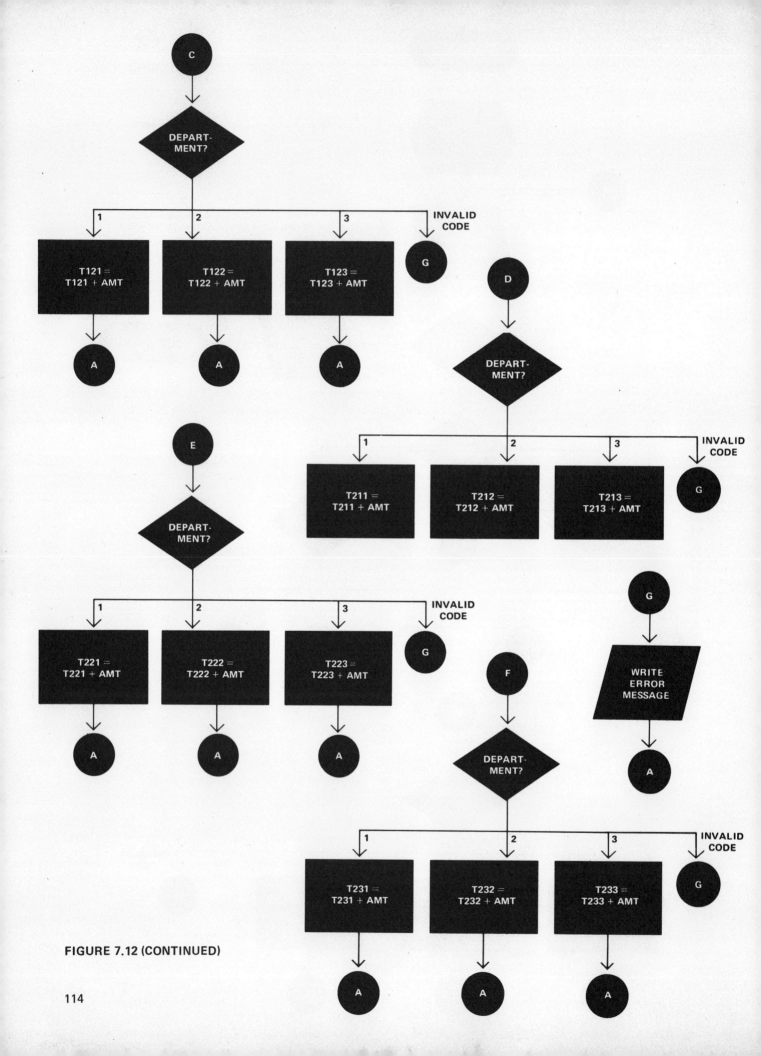

FIGURE 7.12 (CONTINUED)

Exercises

1. What would happen if the totals were initialized each time the loop was executed?
2. The connector symbols directing control to point A represent what kind of building block? Why are they important in the coded program?
3. Expand the flowchart to count the number of returns to each division.
4. Redraw the flowchart so that only two-way conditional branches are used.
5. Change the algorithm so that all the amounts returned are categorized by department only. Total and print out the amounts returned to all departments 1, departments 2, and departments 3.

UNIT 5 PREPARING REPORTS WITH LITERAL TEXT AND VARIABLE DATA

PROBLEM ● Bristol Manufacturing needs a program that will prepare a weekly personnel report. It should list all new and terminated employees by department and request specific information from the department manager. Reports are to be signed and sent back to the personnel department.

The information to be used to generate the report looks like this:

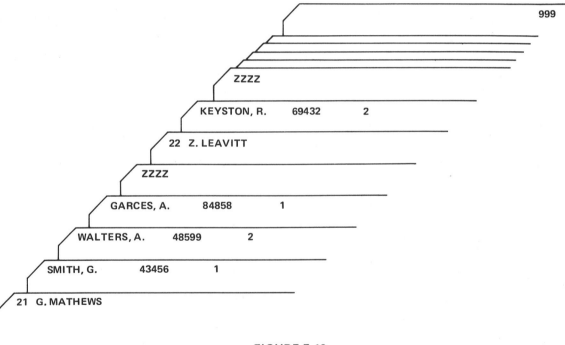

FIGURE 7.13

The first card is the department header listing the department number and manager's name. Next come the employee cards containing a code number and name. This code specifies the status of the employee:

1 = New employee
2 = Terminated employee

A ZZZZ trailer record indicates there are no more records for that department. Each trailer record, except the last, is followed by department. A 999 trailer record indicates the end of the file.

Chapter Seven: Applied Programming Logic

The report for the personnel department looks like this:

```
                    PERSONNEL REPORT
                    DEPARTMENT NO. 21

TEXT A      TO:   G. MATHEWS

            PLEASE REVIEW THE NAMES OF THE EMPLOYEES BELOW.
            FILL IN THE INFORMATION REQUESTED AND RETURN TO
            THE PERSONNEL DEPARTMENT BY NOON ON FRIDAY.

                SMITH, G.    43456

TEXT B              DATE ASSIGNED _____

                    ASSIGNED TO SUPERVISOR _____

                WALTERS, A.   48599

TEXT C              DATE TERMINATED _____

                    REASON _____

                GARCES, A.   84858

TEXT B              DATE ASSIGNED _____

                    ASSIGNED TO SUPERVISOR _____

            TOTAL ASSIGNED TO DEPT.: 02    TOTAL TERMINATED: 01

            THE INFORMATION AND ACTIONS LISTED ABOVE REGARDING
TEXT D      PERSONNEL IN DEPARTMENT 21 HAVE BEEN VERIFIED AND
            APPROVED.

            SIGNED: _____
                                        G. MATHEWS
            DATE: _____
```

FIGURE 7.14

This report has four main text sections. Text A is the opening message and heading. During execution the department number and manager's name are read in and printed out. Next, the report lists an employee's name and identification number followed by either Text B or Text C. Text B, for new employees, asks for the supervisor's name and the assignment. Text C, for terminated employees, requests the reason for termination and the date. Text D, summary data for the department, appears at the end of the report. It shows the total number of employees assigned and terminated and includes provisions for the department manager's signature.

SOLUTION • This program illustrates how to prepare a summary report, forms, or other documents by intermixing literal text and variable data.

Two nested loops read in and process the input file. The outer loop processes the records for each department, and the inner loop processes the individual employee records. A conditional branch determines which of two alternate messages the computer will output regarding each employee. Two counters within the inner loop count the number of new and terminated employees for each department. Two tests for trailer record, each testing for a different condition and field, are used. One tests for 999 in columns 78-80 and terminates the outer loop and program execution. The other tests for ZZZZ in the same field and terminates the inner loop.

BRISTOL MANUFACTURING

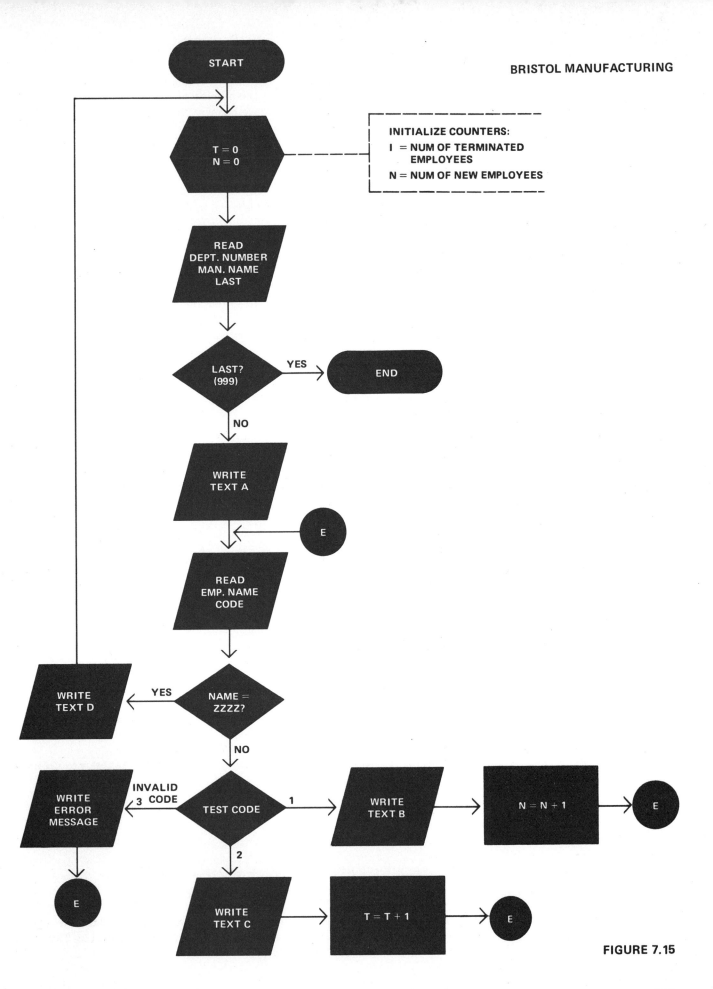

FIGURE 7.15

Chapter Seven: Applied Programming Logic

117

The program begins by entering the outer loop and initializing to zero the fields that count the number of new employees (N) and terminated employees (T). Next the first record, the department header, is read in. It contains the department number and the name of the department manager. The record is tested to see if it has the trailer value in the LAST field. If it does not, the program prints out the portion of the report called Text A. The department number just read in is inserted by the program in the appropriate field as the program outputs the second line. The department manager's name is printed out on the next line. After Text A is printed, the program enters the inner loop. The first employee record is read in. The name field is tested to see if it is ZZZZ, the trailer value. If it is not, the status code field is tested and the program selects one of three paths depending on whether it is one, two, or an invalid code. Path 1 writes Text B, the new employee message, and adds one to the counter N. An unconditional branch sends control back to the beginning of the inner loop to read in the next employee record. Path 2 writes Text C, the terminated employee message, and adds one to the counter T. An unconditional branch returns control back to the beginning of the inner loop to read in the record for the next employee. Path 3 branches to an error routine.

When the ZZZZ trailer is reached, the program branches to write Text D. This summary message intermixes literal text with the tallies from T and N. It also prints the department manager's name and the department number.

An unconditional branch returns control to the beginning of the outer loop to initialize the counters for the next department and the next report. When the 999 trailer is reached, the program terminates execution.

BUILDING BLOCKS USED:

No. 2 — Unconditional branch
No. 3 — Two-way conditional branch
No. 4 — Three-way conditional branch
No. 7 — Counters
No. 9 — Nested loops
No. 10 — Terminating a loop with a trailer record

Exercises

1. Why are the initializing steps included within the outer loop?
2. What would happen if the inner loop returned ahead of the step that reads in the department number and manager's name?
3. Redraw the flowchart using two other means of terminating the loops.
4. Expand the flowchart to test for a third condition — employees on vacation. Structure the program to total the number in this category for each department.
5. Expand the flowchart to total the number of employees in each category for the entire run of the program.

UNIT 6   PROCESSING A DATA FILE WITH ALL RECORDS IN STORAGE AT THE SAME TIME: USING ARRAYS

PROBLEM   ●   The Training Department director of the Saturn Sales Company uses a series of simulations and role-playing exercises to prepare trainees for their jobs. They are rewarded for their skill and improvement in these exercises with one to five points. The director would like a program that will add the points received by the trainees and print out a listing of their totals. She also wants the names of the trainees with the highest number of points marked with three asterisks and those with the lowest number marked with one asterisk. She has 43 trainees in her class and the highest possible score is 30.

The input data is on 43 punched cards. Each card contains a name and six scores:

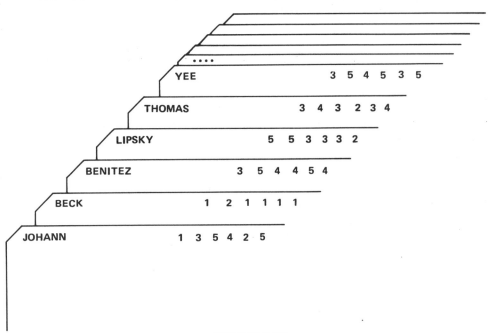

FIGURE 7.16

The listing generated by the program should look like this:

NAME	POINTS
JOHANN	20
*BECK	7
***BENITEZ	25
LIPSKY	21
THOMAS	19
***YEE	25
.....	
.....	

FIGURE 7.17

SOLUTION ● This problem illustrates processing a data file with all records in storage at the same time. This is done by loading the input data into arrays with a limited loop. Sequential limited loops are then used to perform the processing on the arrayed data.

Deciding whether or not to use arrays in a program involves an understanding of several factors. Arrays use up considerable storage space in a computer. This is often a problem when large amounts of data are involved. Whenever possible it is better to process the records in a file singly.

Chapter Seven: Applied Programming Logic

However, since I/O functions are usually slower than central processing time, it may be more efficient to have all elements in a data file accessible to the program at one time. In some procedures (e.g., sorts and searches) this is absolutely necessary. The programmer must weigh these factors when selecting the algorithm for the problem.

In this example, the highest and lowest numbers of points are not known until all records have been processed. The program must be able to go back to the beginning of the file in order to print out asterisks beside the appropriate names. Since the input is on punched cards, the deck would have to be reloaded, which is impractical in many situations. (Had the data been recorded on magnetic tape, instead of being punched into cards, it would be easier for the program to rewind the tape to the beginning of the file before generating the output. This would avoid the necessity of storing all the names in an array.)

After considering that the input is on punched cards and that the data set is relatively small, the programmer has decided to load the information into two parallel arrays for processing. The total in points for each trainee is calculated at the same time.

The program goes back to the beginning of the file to locate the highest and lowest scores. It checks each total against the existing high and low values, replacing these values when indicated. When the entire file has been processed, the two extreme scores are known. The program loops back again to the beginning of the file to prepare the output. Each score is compared to the high and low and asterisks are printed out in the appropriate locations as the listing is generated.

The program begins by reserving 43 storage spaces for NAME and 43 spaces for POINTS. Next HIGH is initialized to one, which is the lowest possible score it could be. LOW is initialized to 43, the highest it could ever be. Then the program enters the first limited loop. The first record is read in and the name is stored in location 1 of the array NAME. The six scores are read in and summed, and the total is stored in the first location of the array POINTS. The program automatically increases the index and loops back to read in and process record number two. After all 43 records have been processed, the loop terminates and control drops to the next sequential limited loop. This loop finds the highest and lowest totals. The index is reinitialized to one, which points the program back to the first record in the file.

Tracing the input data through the operations in this loop demonstrates how the extreme values are found. POINTS(1) holds the value 20 (for Johann). A conditional branch tests 20 against the HIGH value, which is equal to one the first time through the loop. Since POINTS(1) is greater than HIGH, its value becomes the new HIGH. HIGH now equals 20. Another conditional branch compares POINTS(1) to LOW. Since POINTS(1), 20, is less than LOW (initialized to 43) its value becomes the new LOW. An unconditional branch sends control to the last statement in the limited loop, the index increases, and the program returns to the beginning of the loop to process the next record.

POINTS(2) equals 7 (for Beck). This is less than HIGH (20), so POINTS(2) is tested against LOW, which is equal to 20. Since POINTS(2) is less than 20, its value (7) now becomes the new LOW. An unconditional branch sends control to the last statement in the limited loop, the index increases, and the program returns to the beginning of the loop to process the next record.

POINTS(3) is equal to 25, which is greater than the present value of HIGH. The value of HIGH is reset to 25, the program sends control to the end of the loop, the index increases, and the loop repeats. POINTS(4) is now tested. Its value, 21, is less than HIGH (25) and greater than LOW (7). Control takes the third pathway, which drops to the end of the limited loop. The index increases and the loop repeats. This process continues until all 43 records have been tested. At that time, the value of HIGH will represent the highest score achieved by the trainees, and the value of LOW will represent the lowest.

Control now enters the third sequential limited loop, which lists all trainees and their total points, and it marks those with the highest and lowest scores. The index is again initialized to one in order to point the program back to the first record in the file, which is stored in the first locations of the arrays.

POINTS(I) is tested against the final value of HIGH (25) with a conditional branch. If

SATURN SALES COMPANY

START
↓
RESERVE STORAGE SPACES: NAME (43), POINTS (43)
↓
HIGH = 1
↓
LOW = 43
↓
REPEAT STEPS 43 TIMES (I = 1 TO 43)
↓
READ NAME(I), P1, P2, P3, P4, P5, P6
↓
CALCULATE POINTS(I)
POINTS(I) = P1 + P2 + P3 + P4 + P5 + P6
↓
A

FIGURE 7.18

Computer Algorithms and Flowcharting

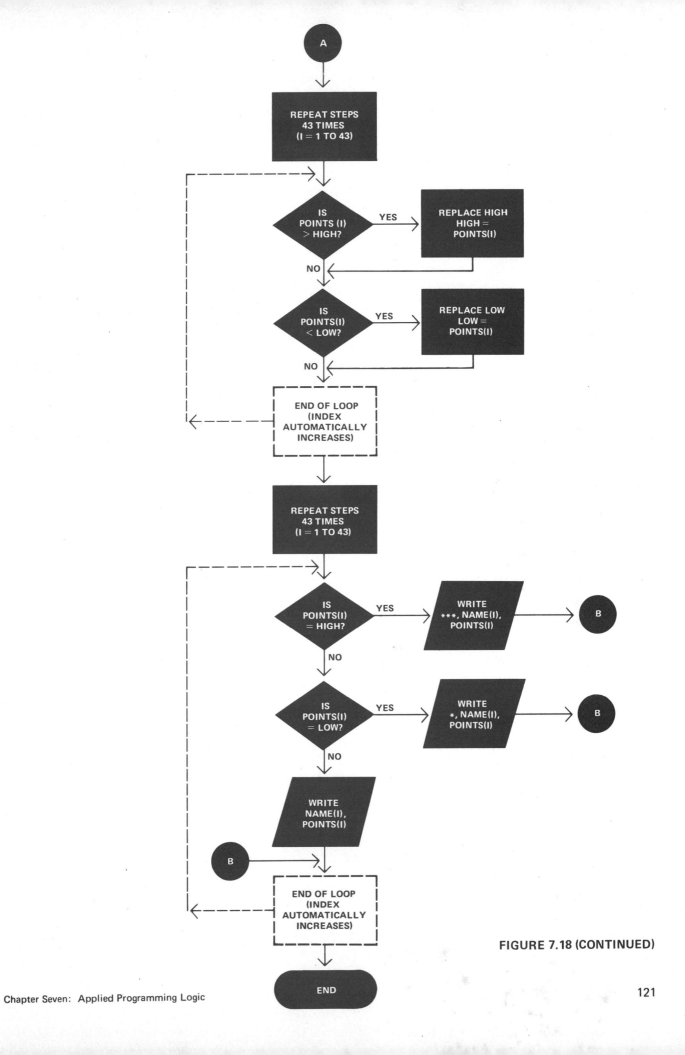

FIGURE 7.18 (CONTINUED)

POINTS(I) is equal to HIGH, the program branches to an output sequence that prints three asterisks, the name of the trainee, and the total points. The index increases by one and control returns to process the data in the next locations of the arrays.

If POINTS(I) is not equal to HIGH, it is tested against the final value of LOW (7). If POINTS(I) is equal to LOW, the program branches to an output sequence that writes one asterisk, the name, and the total score. An unconditional branch sends control to the end of the loop, the index increases, and the next set of data is processed.

If POINTS(I) is not equal to HIGH and not equal to LOW, control branches to a third output sequence that prints out only the name and the score. Control falls through to the end of the loop, the index increases, and the program loops back to test the next value.

This process continues until all 43 lines have been written. Each trainee who tied for HIGH will have three asterisks. All those who tied for LOW will have one asterisk.

Coding Note: The second and third sequential limited loops include two two-way conditional branches within their range. In each case, three alternative pathways are created by the branches. All three pathways must return control to the end of the limited loop in order to automatically increase the index and repeat the loop. Therefore, when coding the instructions, unconditional branches must terminate two of the pathways and send control to the next appropriate point in the loop. The third pathway will automatically fall through to the end of the loop, since instructions are executed sequentially. In both of the loops in this example, the most frequently executed branch was selected to fall through. This saves the execution time of one instruction each time the branch is taken.

BUILDING BLOCKS USED:

No. 2 — Unconditional branch
No. 3 — Two-way conditional branch
No. 8 — Sequential loops
No. 13 — Limited loops
No. 14 — One-dimensional arrays

Exercises

1. What would happen if both HIGH and LOW had been initialized to zero?
2. What would be the major differences in the algorithm if the program were required only to calculate the high and low values, but not print out the asterisks?
3. Redraw the flowchart to use only two limited loops. (Suggestion: Combine the first and second loops.)
4. Change the algorithm to handle an undetermined number of records in the input file.
5. Assume the input file is on magnetic tape that can be rewound by the program. Redraw the flowchart to process the records one at a time, eliminating the arrays.

UNIT 7   USING DECISION TABLES

PROBLEM ● Blythe Clothiers wants a program that will prepare its monthly billing. The program must be able to produce the appropriate messages and enclosures for each account. This means the program must be able to select one or more output sequences depending on the status of the customer's transaction record. Blythe also wants listings of those accounts who show credit balances and those who are past due in their payments.

The accounts receivable file is in alphabetic order and has a varying number of records each month. Each record contains the name of the account, the address, the balance (or credit) due, the credit limit, if any, and five status fields. Each field represents a specific condition. An X punched in the field indicates that the condition is true. Any other character, or a blank in the field, indicates that the condition is untrue. The five status fields are as follows:

S1 = Active account (transactions within the last 3 months)
S2 = Balance due
S3 = Balance past due
S4 = Credit balance
S5 = Credit limit

An input record looks like this:

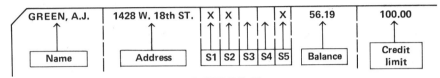

FIGURE 7.19

Blythe wants an advertising letter to go to all customers except those who show a past-due balance. In addition, all customers with a current balance should receive a regular monthly bill. Accounts with a credit balance should receive a message showing the amount. Accounts that have been assigned a credit limit should be informed of the amount still available for use. Past-due accounts should receive a warning letter. The company also wants a report listing the accounts with credit balances, the amounts, and their total. A list of accounts with past-due balances, the amounts, and the total should also be prepared.

BLYTHE CLOTHIERS
DECISION TABLE

		1	2	3	4	5	6	7	8	9	10	11	12
S1	ACTIVE ACCOUNT	Y	Y	Y	Y	Y	Y	Y	Y	N	N	N	N
S2	BALANCE DUE	N	N	N	N	Y	Y	Y	Y	N	N	N	N
S3	BALANCE PAST DUE	N	N	N	N	N	N	Y	Y	N	N	N	N
S4	CREDIT BALANCE	N	N	Y	Y	N	N	N	N	Y	Y	N	N
S5	CREDIT LIMIT	N	Y	N	Y	N	Y	N	Y	N	Y	N	Y
	REGULAR BILLING					X	X						
	WARNING LETTER							X	X				
	LIMIT NOTICE		X		X		X				X		X
	CREDIT MESSAGE			X	X					X	X		
	CATEGORIZED			X	X				X	X	X		
	ADVERTISING LETTER	X	X	X	X	X	X			X	X	X	X

FIGURE 7.20

SOLUTION • This problem illustrates the use of decision tables. The decision table is an invaluable guide in flowcharting a program with many branches. In this example, the program must select one or more of six possible output sequences depending on the values of five variables.

In this program the records in a file are processed both sequentially and as a group. Information from selected records is stored in one of four arrays that are loaded by counters. The arrays are printed out at the end of the program by limited loops. The final tallies of the counters have been used as the limit in the loops to terminate execution.

Preparing the decision table is the first step in solving this problem. The one shown on page 123 graphically illustrates all possible conditions and related actions. This decision table indicates that there are 12 rules, encompassing a variety of actions, that must be incorporated within the program.

Next, the programmer prepares a flowchart. First, 250 storage spaces for each of four arrays are reserved. Two arrays hold the data on the past-due accounts, including the names and amounts past due. The next two arrays hold the data for the accounts with credit balances, the names and amounts in corresponding locations. Then, two counters that load the arrays are initialized to zero. Two running totals that accumulate the past-due balances and the credit balances are initialized to zero.

The decision table indicates that there are 12 rules, encompassing a variety of actions, that must be incorporated within the program.

Next, the programmer prepares a flowchart. First, 250 storage spaces for each of four arrays are reserved. Two arrays hold the data on the past-due accounts, including the names and amounts credited in corresponding locations. Then, two counters that load the arrays are initialized to zero. Two running totals that accumulate the past-due balances and the credit balances are initialized to zero.

The program enters the first loop. The first record is read in and tested to see if it is the end-of-file or trailer record. If it is not, the program prints out the name and address of the customer on the line printer. Then the active and inactive accounts are separated by testing to see if S1 is equal to X. If it is not, the account is inactive and the program branches to print C. If S1 is equal to X, the account is active and the program tests to see if S2 is equal to X. If it is, it indicates there is a balance due. Accounts with no balance due would not have a past-due balance, so the program branches to point C to test S4. S3 is tested on those accounts with balances due to see if they are past due or current. Bills are printed for those with current balances and the program branches to point D.

The names of accounts with past-due balances and the amounts are added to arrays ACCT and PD. (The detailed explanation of this procedure is shown in the flowchart for the routine ACCT-PD, Figure 7.22A.) A warning letter is printed out for these customers. The program branches back to point A to read in and process the next record.

The program has branched to point C for all inactive accounts and for active accounts with no current balances due. S4 is tested to see if the customer has a credit balance. If not, the program branches to test S5. If there is a credit balance, the name of the account and the amount are added to the arrays CUST and CB. (This process is explained in more detail in the flowchart for the routine CUST-CB, Figure 7.22B.) A notice of the amount of the credit balance is printed out.

The sequence at point D tests to see whether there is a limit on the amount of credit available to the customer, calculates the amount of credit left, and writes a message to that effect on the bill or credit notice being output. First, S5 is tested to select those customers with credit limits. Then S4 is tested again to branch to the proper mathematical calculation for those with a balance due. The program subtracts the amount of the balance from the credit limit to reach the new limit (NEWL). For those with credit balances, the program adds the amount of the credit to the limit to calculate the amount of limit available. Then the two branches join to output the new credit limit.

Program flow continues and the advertising letter is prepared for all customers except those with past-due balances. There is a branch back to point A to read the next record.

The flowchart for the routine ACCT-PD shows that the two arrays are loaded with a counter. The name of the counter and the array index that it controls is I which was initialized to zero at the beginning of the program. The first step within the routine is to increase I by one. The name on the record being processed is stored in the location of ACCT having a value of I. The amount of the past-due balance is stored in the corresponding location of PD. The amount of the past-due balance is also added to a running total PDTL.

The next time this routine is executed, I is increased again by one and the name and amount are loaded into the next locations of the arrays. When the record for the last past due customer has been processed, the value of I will be equal to the number of records stored in the arrays.

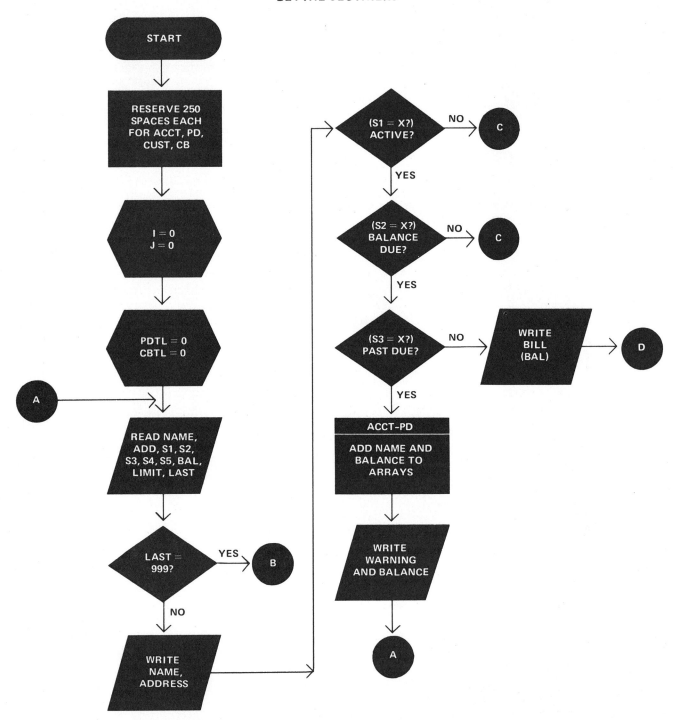

FIGURE 7.21

Chapter Seven: Applied Programming Logic

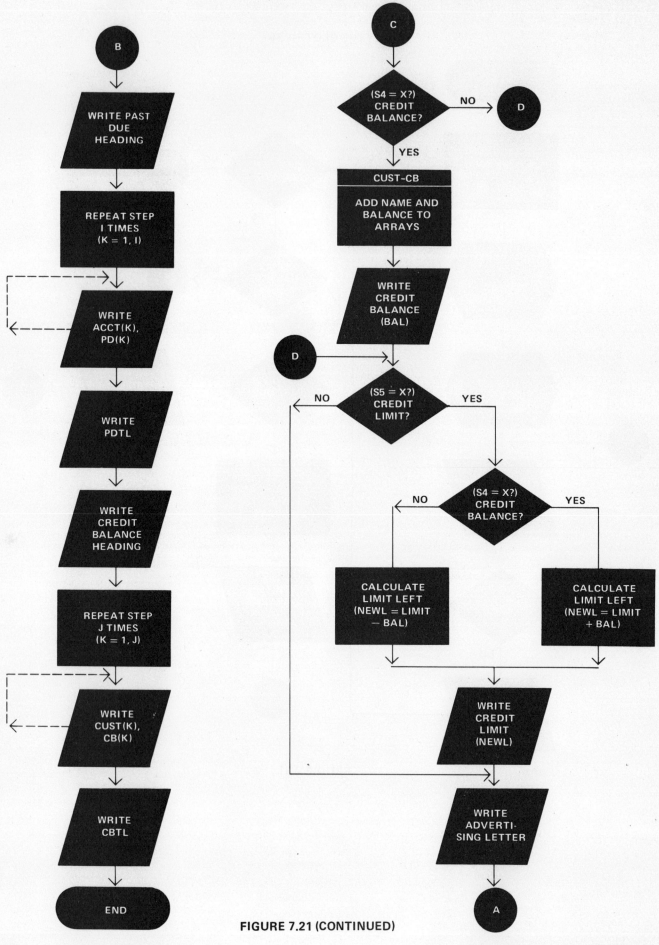

FIGURE 7.21 (CONTINUED)

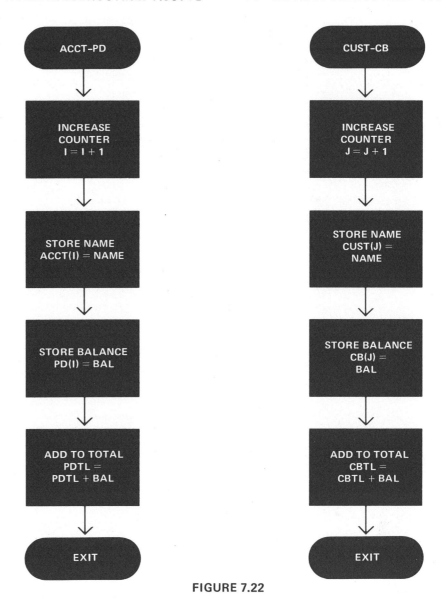

FIGURE 7.22

The routine CUST-CB works in a similar manner. A counter J assigns the subscripts for the arrays CUST and CB. CBTL is a running total of the amounts of the credit balances. When the record for the last customer with a credit balance has been processed, J will be equal to the number of records stored in the arrays.

When the record with the 999 trailer value is reached during the read sequence (point A), the program branches to point B to print out the lists of accounts with past-due and credit balances. The heading for the past-due report is printed, and then the program enters a limited loop that prints out the arrays ACCT and PD. The loop will be repeated I times (I equals the number of records stores in these arrays). Then the value of PDTL, the total amount of past-due balances, is printed out.

Chapter Seven: Applied Programming Logic

Next the heading for the credit balance report is printed. The program enters the last limited loop and prints out the arrays CUST and CB. This loop will be executed J times (J is equal to the number of records stored in these arrays). The value of CBTL, the total amount of the credit balances, is printed out and the program terminates execution.

BUILDING BLOCKS USED:
No. 2 — Unconditional branch
No. 3 — Conditional branch
No. 7 — Counters
No. 8 — Sequential loops
No. 10 — Terminating a loop with a trailer record
No. 13 — Limited loops
No. 14 — Manipulating one-dimensional arrays

Exercises

1. Why were counters, rather than limited loops, used to load the arrays?
2. How could the input record be structured to test all conditions with only one status field?
3. Draw the portion of the flowchart that would test and separate the input data in Exercise 2 into active and inactive accounts.
4. This program first separates accounts into active and inactive accounts. Redraw the flowchart so that the program first separates accounts into those with a balance due and those with no balance due.
5. Expand the program to generate a report listing the accounts with credit limits, the amounts still available, and the total of the amount available.

UNIT 8   PROCESSING MULTIPLE INPUT/OUTPUT FILES

PROBLEM ● Merit Manufacturing has a staff of ten sales representatives who call upon customers in several Eastern cities. Some accounts are located in rural areas and others are concentrated in urban areas. Two sales representatives serve customers at the home site. Each sales representative has a different rate of commission, depending on the distance traveled to reach the accounts. The rates vary from 20% to 40% depending on the route. Merit wants a program that will calculate the weekly payroll for the sales staff and the commission report for the company.

The data on each sale and the representative's code number are punched onto a separate card (see Figure 7.23).

The output report generated by the program should list the amount and number of sales, commission rate, and total commission earned by each representative that week. The last item on the report should be the total commission paid to all ten representatives for the week (see Figure 7.24).

The program should also prepare a paycheck for each representative for the amount of commission earned.

SOLUTION ● This program illustrates processing files and arrays in random as well as sequential order. Two input files are accessed by the program and two output files are generated.

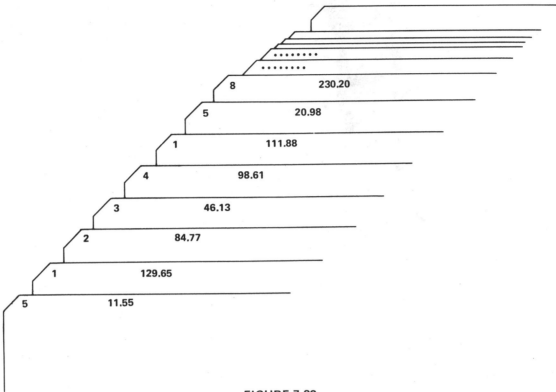

FIGURE 7.23

	COMMISSION SUMMARY			
NAME	TOTAL SALES	NUMBER OF SALES	COMMISSION RATE	COMMISSION
T. SANCHEZ	893.20	9	.40	357.28
M. MCCOY	749.59	12	.35	262.36
L. BREITZ	1422.99	38	.25	355.75
....				
....				
			TOTAL COMMISSIONS PAID	$3,292.44

FIGURE 7.24

The programmer selected the following solution for this problem. Two input files will be used. The first is the punched card file containing the information regarding individual sales. The cards in the file are in random order, and a blank trailer card is used to mark the end-of-file. The second input file is a table containing the name of each representative and the commission rate:

```
T. SANCHEZ   .40
M. MCCOY     .35
L. BREITZ    .25
  . . . . .
  . . . . .
```

Chapter Seven: Applied Programming Logic

This table is recorded on magnetic tape and accessed by the program during execution. (If magnetic tape facilities had not been available, punched cards could have been used instead.)

Two output devices will be used by the program during execution. A card punch will prepare the payroll checks and a line printer will write the commission summary.

The algorithm used to solve this problem utilizes four arrays. One stores running totals that count the number of sales made by each representative; another stores running totals of the amounts of sale. Since the records from which this data is calculated come in random order, an incrementing counter or index would not access the appropriate array locations. But since each representative has a unique number that is punched into each of the sales cards, it is used as the subscript to access the array location that belongs to that salesperson. For example, T. Sanchez is assigned code number 1, and position 1 in all the arrays contains data related to her. L. Breitz has code number 3, and all his data is stored in position 3 of the arrays.

The last two arrays store the data from the name and rate file. These arrays are loaded and manipulated sequentially with limited loops. Sequential limited loops are also used to initialize the running totals (array T and array CT), to calculate the individual commissions and the company total, and to prepare the output files.

The program begins by reserving storage space for the four arrays (ten spaces for each). A limited loop initializes all of the locations in the arrays CT and T to zero. The field that will hold the variable TOTAL is set to zero. The first record from the card file is read in. It contains the code number 5 and the amount 33.55. AMOUNT is tested to see if it is the blank trailer record. If it is not, the code number field is tested to make sure it is between one and ten. If it is not, the program branches to an error routine. If the code is between one and ten, it is used by the program as a subscript to indicate which location in array CT is needed. Each location in CT is a counter that adds up the number of sales made by one representative. Each time a card is read in, the counter in the location corresponding to the code number is incremented by one.

For the first record, position 5 in the array will be incremented by one (CT(5)=0+1). Each position in the array T is a running total of the amounts of the sales made by one representative. In this example, the AMOUNT will be added to position number 5 in the array T (T(5)=0+33.55). An unconditional branch sends the program back to read the next record in the card file.

This record contains the code number 1 and the amount 129.65. Location 1 in the array CT is incremented by one (CT(1)=0+1), and the amount is added to location 1 of the array T (T(1)=0+129.65). The program loops back to read and process the next record. This procedure continues until the end-of-file (blank trailer record) is reached.

At this point, control transfers to a limited loop that reads in the data from the name and rate file. (On many systems the program must include commands at this point to open, prepare, or rewind the magnetic tape files.) The items in this file are read in in pairs. The first item is assigned to location 1 in the array NAME, the second to location 1 in the array RATE. The next two items go into the second locations in each array, and so on, until ten records have been stored in the two arrays.

Then the program writes the headings for the commission summary. A third limited loop calculates the commissions for each representative, using the data stored in the corresponding positions in the arrays T and RATE. When I=3 (for L. Breitz), the total stored in location T(3) will be multiplied by the rate stored in RATE(3), and the result will be stored under COMM. Then COMM will be added to TOTAL (total commissions paid). Next the program will write the data related to representative 3 on the Commission Summary. The name stored in location NAME(3), L. Breitz, will be printed out along with T(3) (the total amount of sales that have been completed), CT (3) (the number of sales), and RATE(3) (the commission percentage) and the commission that has been earned for that week. Last, a check containing the name and amount of commission will be prepared on the card punch. The limited loop then directs the program back to perform the same processing steps for representative 4. When the steps have been performed for all ten representatives the program writes the total and terminates execution.

MERIT MANUFACTURING

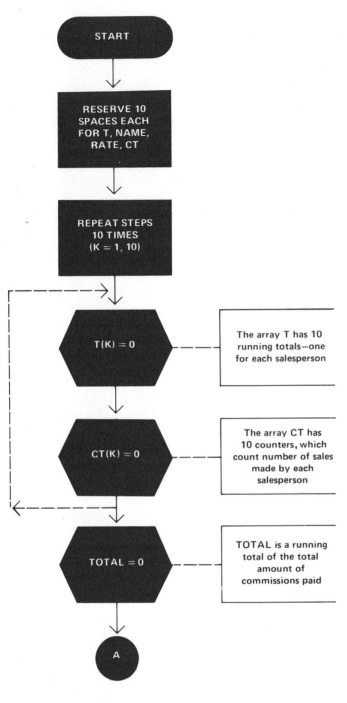
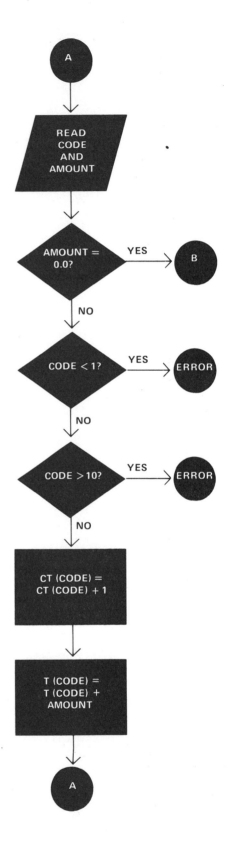

FIGURE 7.25

Chapter Seven: Applied Programming Logic

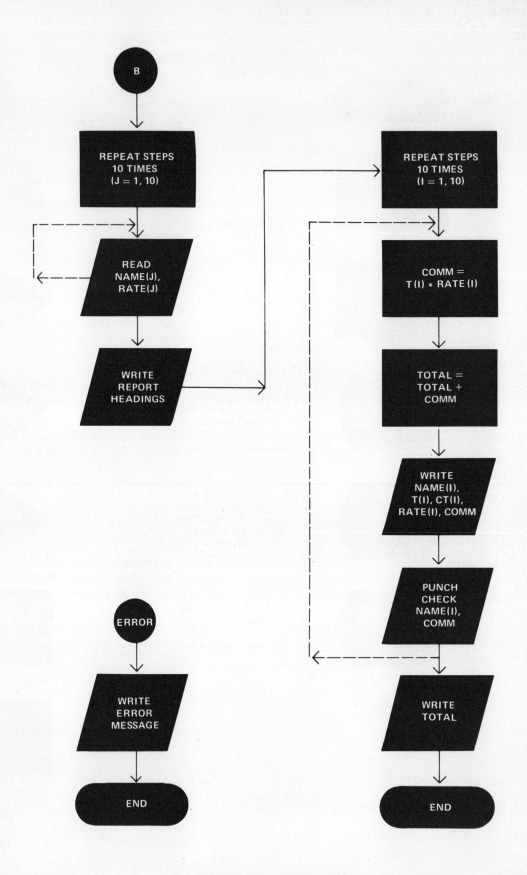

FIGURE 7.25 (CONTINUED)

If an error had been detected in the input card file, the program would have printed an error message and terminated execution. In situations where a large number of input cards are involved, or computer processing takes place at a remote location, it is more advisable to have the program go to the error routine to identify the incorrect input record and then branch back to continue processing the next record.

Note that different subscript names are used throughout the program. The same name could have been used instead. In sequential processing, either method is correct, as long as the names of the index and the subscript it controls are the same.

BUILDING BLOCKS USED:

No. 2 — Unconditional branch
No. 3 — Conditional branch
No. 7 — Counters
No. 8 — Sequential loops
No. 10 — Terminating a loop with a trailer record
No. 13 — Limited loops
No. 14 — Manipulating one-dimensional arrays

Exercises

1. Why was it necessary to initialize the arrays T and CT to zero?
2. What is meant by "processing arrays in random order"? Can this be done on all media?
3. How could the program algorithm have been designed to permit sequential access of array locations?
4. Redraw the flowchart so that the program can handle 15 representatives.
5. Suppose the data for the commission table were arranged so that the names were in one file and the rates in a second file. Restructure the flowchart at point B to read in the two files and assign to the appropriate locations the corresponding subscripts.
6. Assume each representative also receives a minimum weekly salary that is added to the commission to calculate the total amount of each paycheck. Expand the flowchart so that the program reads in the weekly salary and computes both the final amount for each representative and the total amount of the payroll.

UNIT 9 PROGRAMMING MATHEMATICAL FORMULAS

PROBLEM ● Zeplin Trailer Company is ready to introduce its new line of recreational trailers and is planning the advertising budget for the sales push. Zeplin wants to determine whether there is a relationship between the amount spent on advertising and the number of sales. During the past 24 months, the company has kept careful records showing the amount of money spent on advertising and the monthly sales. Zeplin wants a program that will determine the effect of each advertising budget on sales.

A statistical technique, correlation analysis, will be used. *Correlation analysis* determines whether there is a relationship between two sets of data and what the direction and the strength of that relationship are. This relationship is called the *coefficient of correlation* and is expressed on a scale of -1 to +1. A coefficient of -1 would indicate a strong negative correlation — the two sets of data are in inverse proportion to one another (the hotter the weather, the less heating is used in homes). A coefficient of zero would indicate that absolutely no relationship exists — there is no observable change in one set of data relative to a change in the other. A coefficient of +1 would indicate a strong positive correlation — the sets of data are in direct proportion (the hotter the weather, the more air conditioning is used in homes).

The programmer's task will be to find the coefficient of correlation for the relationship between the amount of money spent on advertising and the number of sales. To do this, it will be necessary to find a way to express a statistical formula as steps suitable for computer programming.

Zeplin has prepared the input data for this problem on punched cards. Each contains two numbers, the amount of money spent on advertising each month and the number of units sold during that period. There are 24 cards, each representing one month.

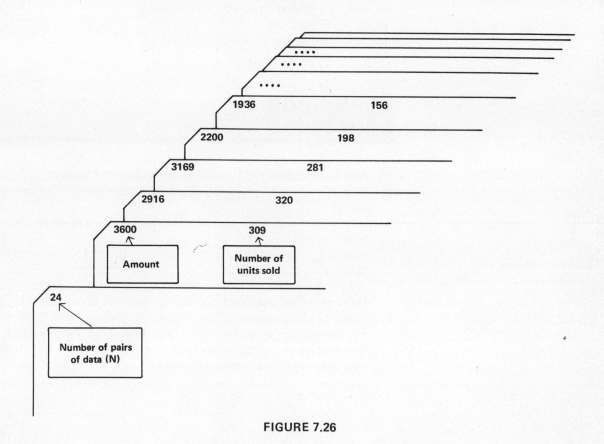

FIGURE 7.26

SOLUTION • This problem illustrates how to express a statistical or mathematical formula as steps suitable for programming. In this example the following formula for correlation analysis is used:

$$r = \frac{n(\Sigma XY) - (\Sigma X)(\Sigma Y)}{\sqrt{[n(\Sigma X^2) - (\Sigma X)^2][n(\Sigma Y^2) - (\Sigma Y)^2]}}$$

The first step in programming this formula is to assign appropriate names to the terms in the formula. (These names will vary according to the computer language being used.)

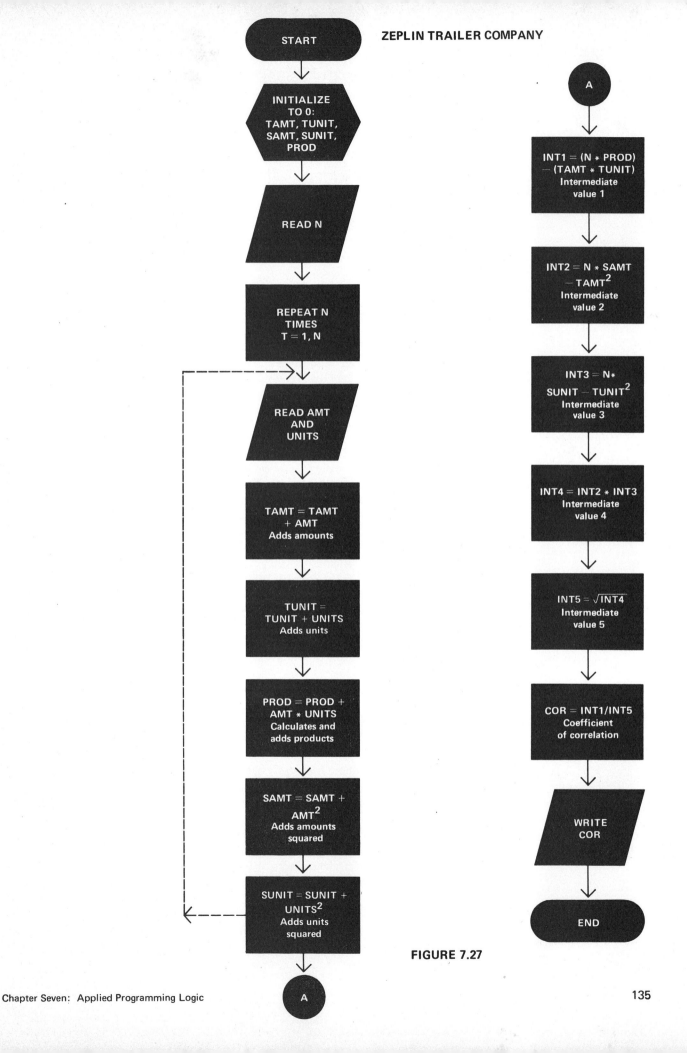

FIGURE 7.27

Chapter Seven: Applied Programming Logic

Here are names selected to represent each term and its meaning:

NAME IN PROGRAM	NAME IN FORMULA	MEANING
N	n	Number of pairs of data
	Σ	Statistical symbol for "sum of"
AMT	X	Amount spent per month (first set of data)
UNITS	Y	Number of units sold (second set of data)
TAMT	ΣX	Running total for AMT
TUNIT	ΣY	Running total for UNITS
PROD	ΣXY	Sum of products — each pair of data is multiplied (AMT * UNITS) and added to running total.
SAMT	ΣX^2	Sum of squares — each value of AMT is squared and added to running total.
SUNIT	ΣY^2	Sum of squares — each value of UNITS is squared and added to running total.
COR	r	Coefficient of correlation

Next, these names are substituted in the formula:

$$COR = \frac{(N * PROD) - (TAMT * TUNIT)}{\sqrt{((N * SAMT) - TAMT^2) * ((N * SUNIT) - TUNIT^2)}}$$

Now the programmer is ready to program the mathematical steps in the formula according to the rules and hierarchy of computer programming.

Briefly, the algorithm followed in the program is this: A limited loop is used to read in the data file. The first card in the file contains a value named N. N indicates the number of cards (or pairs of data) that are present in that particular file, and it is used to limit the number of times the loop is executed. By changing this value, the program can process data files of different lengths.

The other records in the data file contain two numbers each. As each pair is read in, several accumulative mathematical operations are performed on the data. After the loop has terminated, several more mathematical operations are performed on the totals to complete the calculations. Finally, the coefficient of correlation is printed out.

The program begins by initializing to zero the mathematical fields that are used to accumulate running totals. Next the current value of N (in this case, 24) is read in. Then the program enters a limited loop. The loop reads in the first pair of data (AMT and UNITS) from one record. Each value is added to a running total (TAMT and TUNIT). AMT and UNITS are multiplied and their product is added to a running total (PROD). Finally, each value (AMT and UNITS) is squared, and these values are added to running totals (SAMT and SUNIT). Then the program loops back to perform the same operations on the next pair of data.

After N (24) pairs have been processed, the program leaves the loop and performs the rest of the operations indicated by the formula. Several intermediate values are produced, and one function, square root, is called in. Finally COR, the coefficient of correlation, is calculated.

The last step in the program outputs the value of COR. Had the programmer wished, a sequence could have been added to the program to print out the appropriate interpretation of the correlation score.

The number of mathematical operations performed in each programming step may be varied to suit the programmer's familiarity with mathematics and the computer language being used. Using fewer steps may be more efficient for some compilers and may require shorter execution time, but it can make debugging and documentation of programs more difficult.

BUILDING BLOCKS USED:
No. 1 — Single pass execution
No. 13 — Limited loops

Exercises

1. What changes would be required in this program if the input data file contained 50 records? Why?
2. If a program such as this were giving incorrect results, where would you recommend inserting temporary WRITE statements as a means of program debugging?
3. Design the read loop to use a different method of handling an undetermined number of records.
4. Expand the flowchart so that the program prints out the appropriate interpretations of the coefficient of correlation.
5. Flowchart the steps that will solve the following formula:

$$ANS = \frac{(A+B)C}{\sqrt{D + A*8}}$$

UNIT 10 INTERACTIVE PROGRAMMING TECHNIQUES

PROBLEM ● Penucci Finance Company wants a program that will compute the total amount of interest due on a given loan. The program is to be stored online at a computer center and accessed through a computer terminal at the loan company's office. Penucci wants the program structured so the company can select one or more options during execution. It must be able to calculate simple interest on the declining balance and installment interest on the full balance.

The program must be able to request and accept the data needed for processing from the remote terminal. The user must have the option of obtaining a printout of a table showing payments and balances for each period during the life of the loan.

SOLUTION ● To solve this problem, the programmer has selected an algorithm involving interactive programming techniques or techniques that allow for a question-and-answer approach. Interactive programming offers the user many advantages. Execution is online and real time. *Online* means that variable data can be entered from the terminal keyboard, which is connected directly to the computer. *Real time* means that the program is executed immediately and results are returned at once. New data can be entered, the program can be repeated, and the new results can be printed out immediately.

Interactive programming adds flexibility to a program. The user can direct the program to carry out some or all of the procedures coded in the program while it is being executed. All or part of the program can be repeated on new sets of data as needed.

The logic followed in designing an interactive program is basically the same as that used in batch programming. In *batch programming* a job is submitted as a complete unit for computer processing. The major difference is the ability to select options *during* execution. The computer waits for a value to be entered from the terminal, compares it against the test condition in a conditional branch, and selects the appropriate pathway, depending on the results of the test.

In this program, the user has the option of calculating simple interest on the declining balance or installment interest on the full balance. The program offers the option of calculating and printing out the monthly payment or calculating and printing out a table showing payments and balances for the life of the loan. Finally, the user can either repeat the entire program on another set of data or terminate execution.

Below is an example of the input and output generated at the terminal when executing an interactive program of this type:

```
PLEASE ENTER SIM TO CALCULATE
   SIMPLE INTEREST ON DECLINING
   BALANCE.  ENTER INST FOR INSTALL-
   MENT INTEREST ON THE FULL
   BALANCE.                                    (PRINTED BY COMPUTER)

? SIM                                          (TYPED BY USER)

PLEASE ENTER PRINCIPAL, INTEREST
   RATE AND TIME (IN MONTHS)                   (PRINTED BY COMPUTER)

? 100.00, 12, 6                                (TYPED BY USER)

THE INTEREST IS $3.53
THE TOTAL AMOUNT TO BE REPAID IS
   $103.53                                     (PRINTED BY COMPUTER)

DO YOU WANT THE MONTHLY PAYMENT?               (PRINTED BY COMPUTER)

? YES                                          (TYPED BY USER)

THE PAYMENT IS $17.26                          (PRINTED BY COMPUTER)

DO YOU WANT A TABLE?                           (PRINTED BY COMPUTER)

? YES                                          (TYPED BY USER)
```

****

SIMPLE INTEREST ON THE DECLINING BALANCE

*   *   *

PRINCIPAL = $100.00   INTEREST RATE = 12%
LENGTH OF LOAN = 6 MONTHS   MONTHLY PAYMENT = $17.26

PERIOD	APPLIED TO INTEREST	APPLIED TO PRINCIPAL	PAYMENT	BALANCE
1	1.00	16.26	17.26	83.74
2	.84	16.42	17.26	67.32
3	.67	16.59	17.26	50.73
4	.51	16.75	17.26	33.98
5	.34	16.92	17.26	17.06
6	.17	17.06	17.23	0.00

DO YOU WANT TO REPEAT THE PROGRAM?	(PRINTED BY COMPUTER)
? YES	(TYPED BY USER)
PLEASE ENTER SIM TO CALCULATE SIMPLE INTEREST ON DECLINING BALANCE. ENTER INST FOR INSTALL- MENT INTEREST ON THE FULL BALANCE.	(PRINTED BY COMPUTER)
? INST	(TYPED BY USER)
PLEASE ENTER PRINCIPAL, INTEREST RATE AND TIME (IN MONTHS)	(PRINTED BY COMPUTER)
? 500.00, 7, 12	(TYPED BY USER)
THE INTEREST IS $35.00 THE TOTAL AMOUNT TO BE REPAID IS $535.00	(PRINTED BY COMPUTER)
DO YOU WANT THE MONTHLY PAYMENT?	(PRINTED BY COMPUTER)
? NO	(TYPED BY USER)
DO YOU WANT A TABLE?	(PRINTED BY COMPUTER)
? YES	(TYPED BY USER)

****

INSTALLMENT INTEREST ON FULL BALANCE

*   *   *

PRINCIPAL = $500.00 INTEREST RATE = 7%
LENGTH OF LOAN = 12 MONTHS MONTHLY PAYMENT = $44.58

INSTALLMENT	PAYMENT	BALANCE
1	44.58	490.42
2	44.58	445.84
3	44.58	401.26
4	44.58	356.68
5	44.58	312.10
6	44.58	267.52
7	44.58	222.94
8	44.58	178.36
9	44.58	133.78
10	44.58	89.20
11	44.58	44.62
12	44.62	0.00

DO YOU WANT TO REPEAT THE PROGRAM?	(PRINTED BY COMPUTER)
? NO	(TYPED BY USER)

Chapter Seven: Applied Programming Logic

Basically the program uses an algorithm involving several two-way conditional branches and three limited loops. The loops are limited by one of the variables input by the user during execution of the program. Four striped symbols indicate procedures that are described in more detail in separate flowcharts.

The program begins by asking the user to indicate which form of interest is wanted (Input X$), and then to input the data necessary for the calculations. Next the program branches to one of two possible paths, depending on whether the user selected SIM (simple interest) or INST (installment interest).

The branch to INST goes to a routine called INSTINT, which calculates and prints out the interest and total amount to be repaid. The amount of interest is calculated with the formula I=PRT. R is defined for this program as the rate expressed as a decimal, rather than as a percentage. T is the number of months expressed as a fraction of one year. The total loan is calculated by adding the interest to the principal.

SIM branches the program to a routine called SIMINT, which calculates and prints out the interest and total amount to be repaid when the interest is figured on the declining balance. First, the interest to be earned is estimated with the formula I=.6(PRT). The estimated interest is added to the principal, giving an approximate of the total amount to be repaid. The approximate payments are calculated by dividing the amount of the loan by the number of months. Then the program uses these figures as a basis to calculate the exact interest and total amount to be repaid.

The program enters a limited loop that will be executed one time for each month of the life of the loan. During each repetition, the amount of interest earned on the current balance (MTH) is calculated and subtracted from the estimated payment. The balance of the payment (APRIN) reduces the principal (BAL). The interest earned (MTH) is added to TINT. When the loop terminates, TINT gives the exact amount of interest due. TINT is added to PRIN to calculate the exact amount to be repaid. These values are then written out.

Next, (at Point B) the program offers the user the option of calculating the monthly payments for the loan (Input A$). An answer of YES branches the program to calculate and print out this amount. Both paths join again to give the user the option of listing the payment table (Input T$). A NO answer sends the program to point D. A YES answer directs the program to calculate the monthly payments and branch to the appropriate routine.

INSTTAB writes the initial information and headings and then enters a limited loop. The loop will be executed "TIME - 1" times. (For example, if TIME is six months, the loop will be executed five times.) The loop calculates and writes out the installment number (K), the amount of the payment (PAY), and the balance still owing on the loan (LOAN). Then the program writes out information for the last payment, which may be

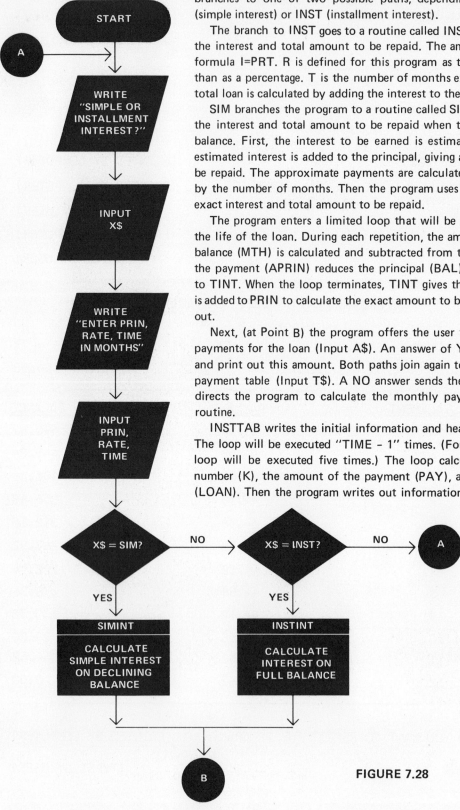

FIGURE 7.28

Computer Algorithms and Flowcharting

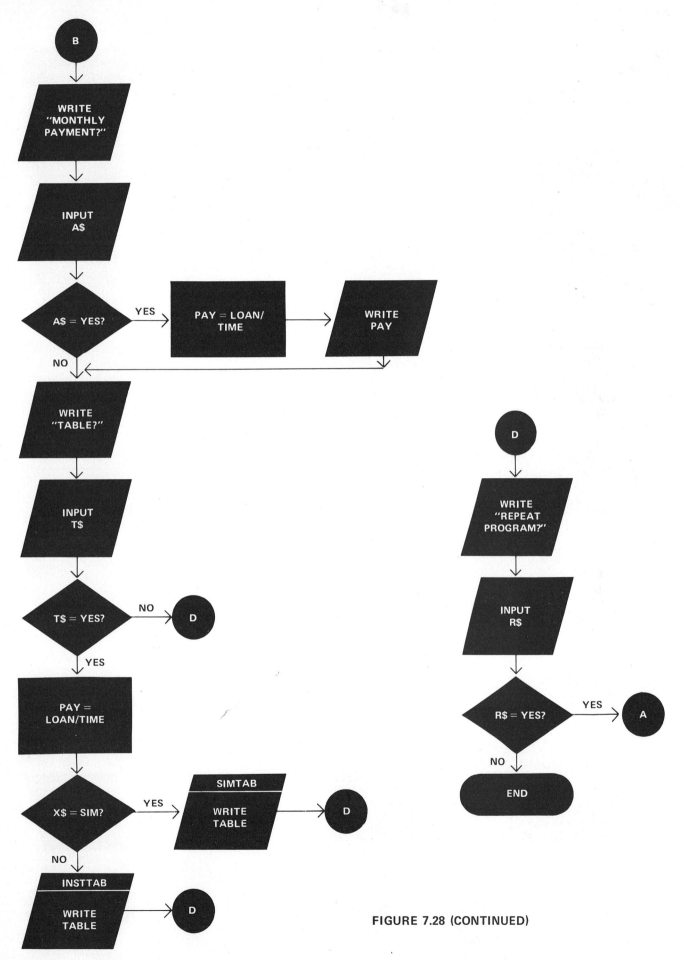

FIGURE 7.28 (CONTINUED)

Chapter Seven: Applied Programming Logic

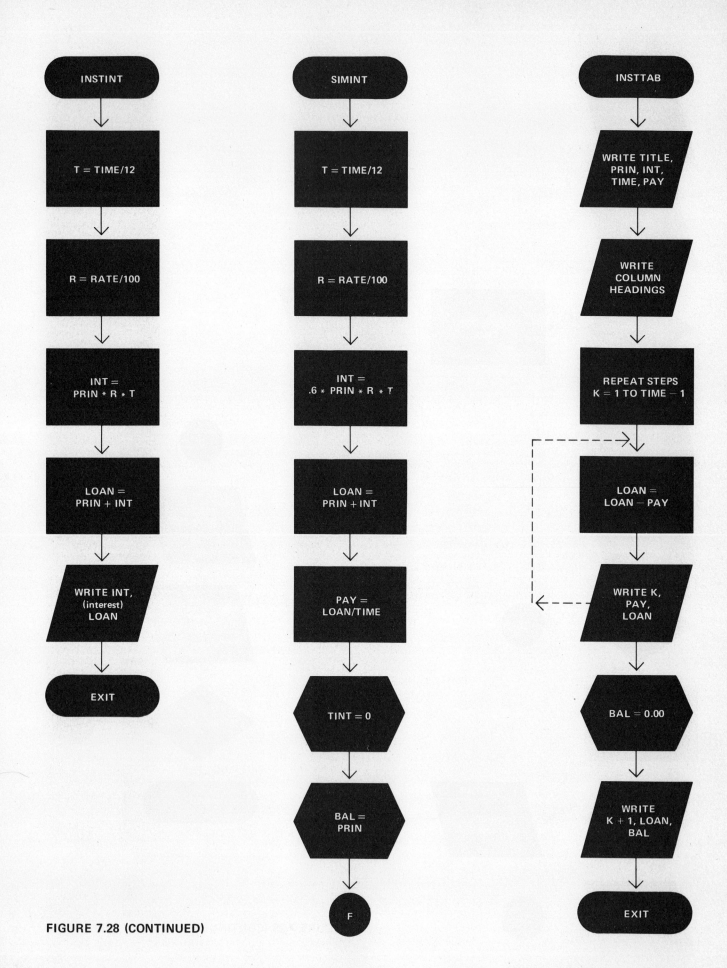

FIGURE 7.28 (CONTINUED)

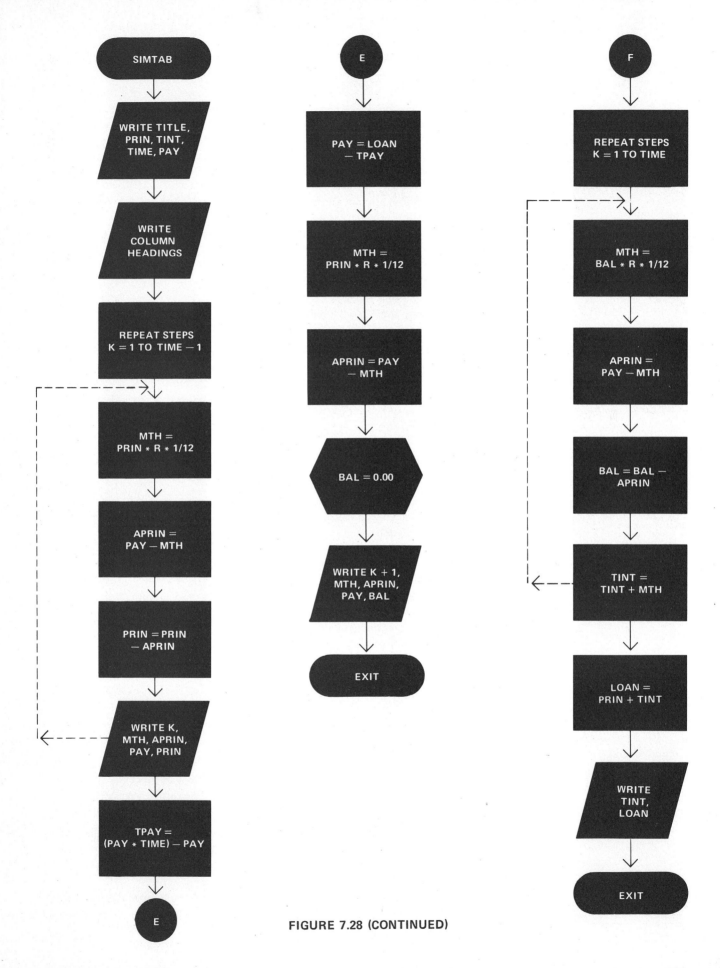

FIGURE 7.28 (CONTINUED)

Chapter Seven: Applied Programming Logic

slightly different from the others. The current amount of LOAN is used as this payment, and reduces the balance still owing to zero.

The routine SIMTAB follows a similar procedure, except for the formula used to calculate the values. After the headings are written, the program enters a limited loop that is executed "TIME − 1" times. The steps are similar to those in the SIMINT routine, but the figures differ slightly since the exact amount of the payments is now known. (SIMINT used an approximate payment.) When the loop terminates, the program calculates and writes the amount of the last payment, which may be slightly different from the monthly ones.

Both routines join at point D. Here the user is offered the last option — whether or not to repeat the entire program on another set of data (Input R$). An answer of YES sends control back to point A at the beginning of the program. An answer of NO terminates execution.

BUILDING BLOCKS USED:
No. 2 — Unconditional branch
No. 3 — Two-way conditional branch
No. 4 — Counters
No. 6 — Simple loop
No. 13 — Limited loops

Exercises

1. Why was the value of PRIN assigned to BAL in the routine SIMINT?
2. Why is it necessary to calculate the monthly payment after A$ is tested and again after T$ is tested?
3. Expand the flowchart to offer the option of generating a monthly statement.
4. Redraw the flowchart so that the program terminates execution if the user enters END at any decision point.
5. Flowchart an interactive program that calculates the monthly payment on a revolving charge account. The monthly interest is 1½% of the unpaid balance. The new balance is calculated by adding the old balance, charges, interest, and subtracting credits and last month's payment, if any.

UNIT 11 PREPARING GRAPHS

PROBLEM • Vaniada Company operates a dozen assembly plants throughout the United States. Some employees in each plant are employed on the assembly line and are paid on an hourly basis. The other employees manage and supervise the line and receive a flat monthly salary.

Vaniada wants a computer program that will print out a graph showing the number of hourly and salaried employees at each plant. This graph will be used by the personnel department for planning future employment needs.

A file of punched cards holds the data to be used by the program. The information for each plant is recorded on a separate card:

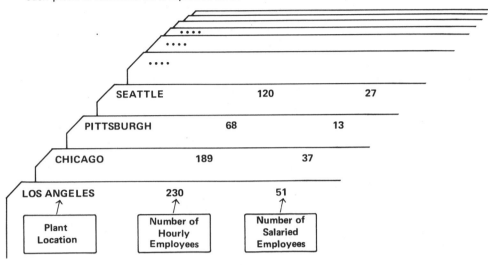

FIGURE 7.29

Vaniada wants the program to generate the personnel graph shown below.

```
                    VANIADA COMPANY
                    PERSONNEL REPORT

        X = HOURLY EMPLOYEES     O = SALARIED EMPLOYEES
            EACH SYMBOL REPRESENTS 5 EMPLOYEES

PLANT
LOCATION     0        50       100      150      200      250
             .         .        .        .        .        .
LOS ANGELES  . XXXXXXXXXXXXXXXXXXXXXXXXXXXXXXXXXXXXXXXXXXXXXX
             . OOOOOOOOOO
CHICAGO      . XXXXXXXXXXXXXXXXXXXXXXXXXXXXXXXXXXXXXX
             . OOOOOOO
PITTSBURGH   . XXXXXXXXXXXXX
             . OOO
SEATTLE      . XXXXXXXXXXXXXXXXXXXXXXXX
             . OOOOO
  . . . . .  . XXXXXXXXXXXXXXXXXXXXXXXXXXXXXX
             . OOOOOOOOO
```

FIGURE 7.30

SOLUTION ● This program illustrates how to prepare one form of bar graph on the computer. It shows how to combine the computer's mathematical and graphic capabilities to conveniently generate pictorial output from numeric data input.

The program processes records in the file singly and in sequence. The algorithm uses two sequential limited loops nested within one outer loop. The outer loop is executed 12 times, once for each of the 12 plants operated by Vaniada. The first sequential inner loop prints out the bar for hourly employees, and the second inner loop prints out the bar for salaried employees. The limits of these loops are determined by mathematical calculations performed on the input data.

VANIADA COMPANY

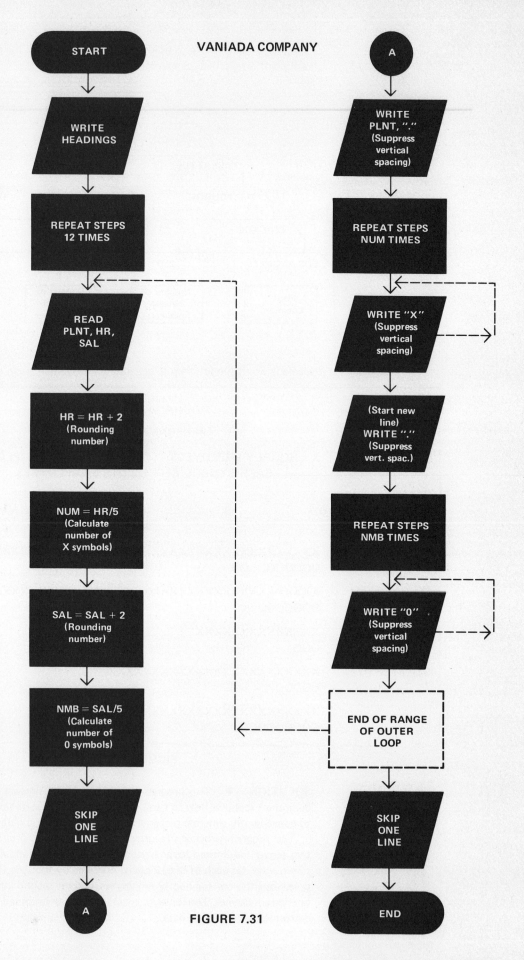

FIGURE 7.31

146

Computer Algorithms and Flowcharting

The algorithm relies on the programmer's ability to control the horizontal and vertical spacing on the line printer output from instructions in the program. Appropriate carriage control characters for the language being employed are used in these WRITE instructions to direct the computer to suppress vertical spacing when they are executed. This has the effect of printing out the data from more than one WRITE statement on the same line. Therefore, the same coded instructions can be used to conveniently output rows with a different number of symbols by varying the number of times they are executed.

Carriage control techniques vary with the language used, but the principle remains the same. In FORTRAN, the carriage control character '+' is used in the FORMAT statement to suppress spacing. In BASIC, a semicolon is used after each item (except the last) in a PRINT statement. In COBOL, the carriage control is indicated in the print line description in the Data Division.

The program begins by writing the headings for the graph. Then it enters a limited loop that is executed 12 times, once for each record in the file. The first record is read in and a constant (2) is added to the number of hourly employees (HR). This will have the effect of rounding up the number if it is equal to, or greater than 3 or 8, when the next calculation is performed. The next step divides HR by five to calculate the number of symbols that should be printed out for each plant (NUM). Each symbol will represent five hourly employees. Then the number of symbols to be printed out for the salaried employees is calculated in the next two steps (NMB). This means that the bar for Los Angeles with 230 hourly employees will be represented by 46 symbols (230+2=232; 232/5=46). The salaried employees will be represented by ten symbols (51+2=53; 53/5=10).

Now the program directs the computer to skip one line and write out the name of the plant and a symbol (.). To suppress vertical spacing, the computer is directed not to move to the next print line. The computer enters the first inner loop, which contains a WRITE instruction for a symbol that will suppress vertical spacing. It will be executed NUM times — once for each symbol (46 times for the Los Angeles hourly employees). All symbols will be on one line. After this loop terminates, the program reaches another write statement that directs the computer to move to the next print line and output a symbol (.). Next the program enters the second inner loop that prints out the symbols for the salaried employees. This will be executed NMB times — once for each symbol on the line (10 times for the Los Angeles salaried employees).

When this loop terminates, the program has completed the first execution of the outer limited loop, and it branches back to read in and process the next record. After the outer loop has been executed 12 times, it terminates. The program branches to a statement that skips one line before terminating execution.

BUILDING BLOCKS USED:
No. 8 — Sequential loops
No. 9 — Nested loops
No. 13 — Limited loops

Exercises

1. What would happen in the sequential loop that prints out the symbols if the vertical spacing were not suppressed?
2. Why were the input values manipulated before being used to control the number of symbols input? Is this always necessary?
3. Redo the flowchart so that the program can handle a number of specified records.
4. Expand the flowchart to print out a third row of symbols to represent part-time employees for each plant.
5. Prepare a flowchart for a program that prints out a bar graph showing the proportion of chemicals in each of six formulas. Assume the formulas are composed of differing proportions of the same three chemicals.

UNIT 12 PERFORMING NUMERIC AND ALPHABETIC SORTS

PROGRAM ● Newton Merchandising Company prepares several mail-order catalogs each year. Hundreds of retail items are described in each catalogue and indexed by both numeric and alphabetic order. Newton wants a program to facilitate preparation of the indexes for these catalogues.

The information describing each item to be included in a catalogue is punched onto a card.

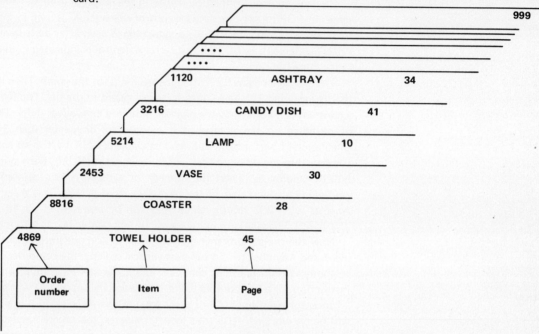

FIGURE 7.32

These cards for a catalogue are grouped together in random order. To prepare the indexes, these cards are first sorted into numeric order by order number, for the first index, and then into alphabetic order by item for the second index.

The indexes generated by the program should look like this:

ORDER NUMBER INDEX		
ORDER NUMBER	ITEM	DESCRIPTION ON PAGE
1120	ASHTRAY	34
2453	VASE	30
3216	CANDY DISH	41
4869	TOWEL HOLDER	45
5214	LAMP	10
8816	COASTER	28
.....		
.....		

FIGURE 7.33

	ALPHABETIC INDEX	
ITEM	ORDER NUMBER	DESCRIPTION ON PAGE
ASHTRAY	1120	34
CANDY DISH	3216	41
COASTER	8816	28
LAMP	5214	10
TOWEL HOLDER	4869	45
VASE	2453	30
. . . . .		
. . . . .		

FIGURE 7.34

Each index should be titled and each column labeled. There should be no more than 25 lines per page.

SOLUTION ● This problem illustrates the type of algorithm that is used to reorder data into a different sequence. In this example a file of records is first rearranged numerically, using the order number as the sort field, and then rearranged alphabetically, with the name of the item as the sort field. The same algorithm is used to perform both sorts.

The program is designed to process a file containing a maximum of 1000 records. Larger files can be manipulated by increasing the number of spaces reserved for each array. The algorithm in this example requires all records to be in primary storage (the main processing section of the computer) at the same time (stored in arrays). The data related to each item in the catalogue is stored in three parallel arrays — ITEM, NO (order number), and PG (page).

The sort procedure is based on a process of comparing the values stored in two adjacent locations of an array and moving the larger value to the lower position. Nested loops are used to perform the sort procedure. The input data is read in and arrayed by a loop controlled by a counter and terminated by a test for a trailer record. Two limited loops are used to output the indexes containing the sorted data.

The program begins by reserving 1000 storage spaces each for the order number (NO), the item name (ITEM), and the page number of the description (PG). Next the counter CT is initialized to one. The initial value is one, rather than zero, since the value of CT is also used to assigned array subscripts.

The program now enters the input loop. The first record is read in and tested to see if it is the end-of-file. If it is not, the three fields are assigned to the first locations of each of the three arrays. The counter is increased by one and the next record is read in and assigned to the second location. This process continues until the end of the file is reached.

Due to the placement and initializing of the counter, the trailer record will have been counted by CT. Therefore, the next step is to decrease CT by one so that it will be equal to the actual number of records to be processed. Next C and KT are assigned the same values as CT. These will be used to limit loops in the sorting and output procedures. Three different names are needed, since their values will be changed during execution.

Then the program begins the first sort routine, SORTNO, which rearranges the records numerically by order number. The steps in the sort procedure will be explained later.

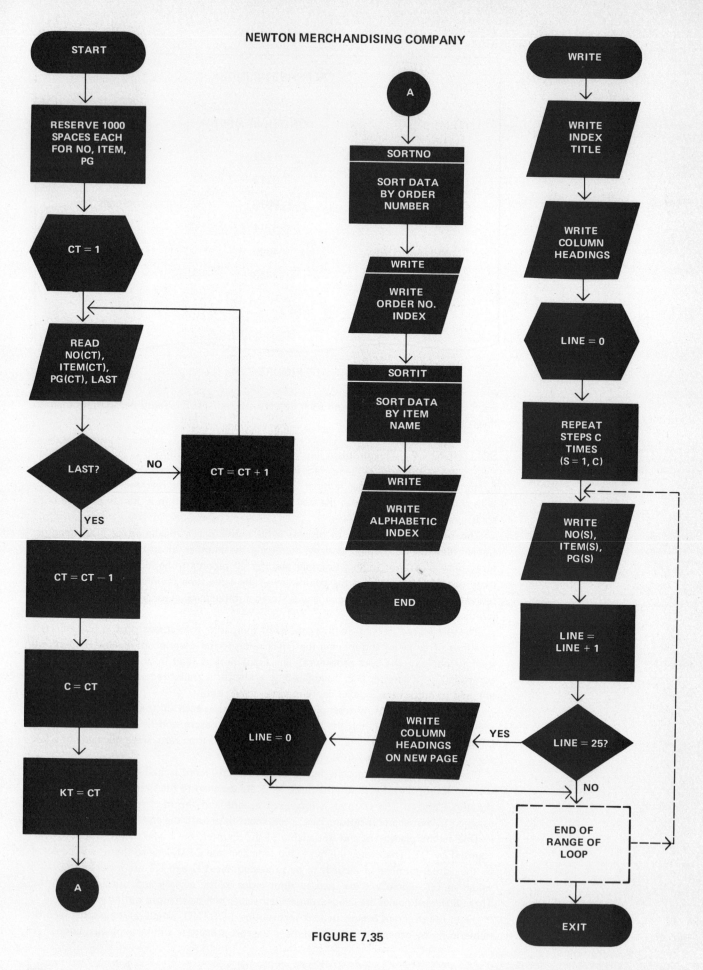

FIGURE 7.35

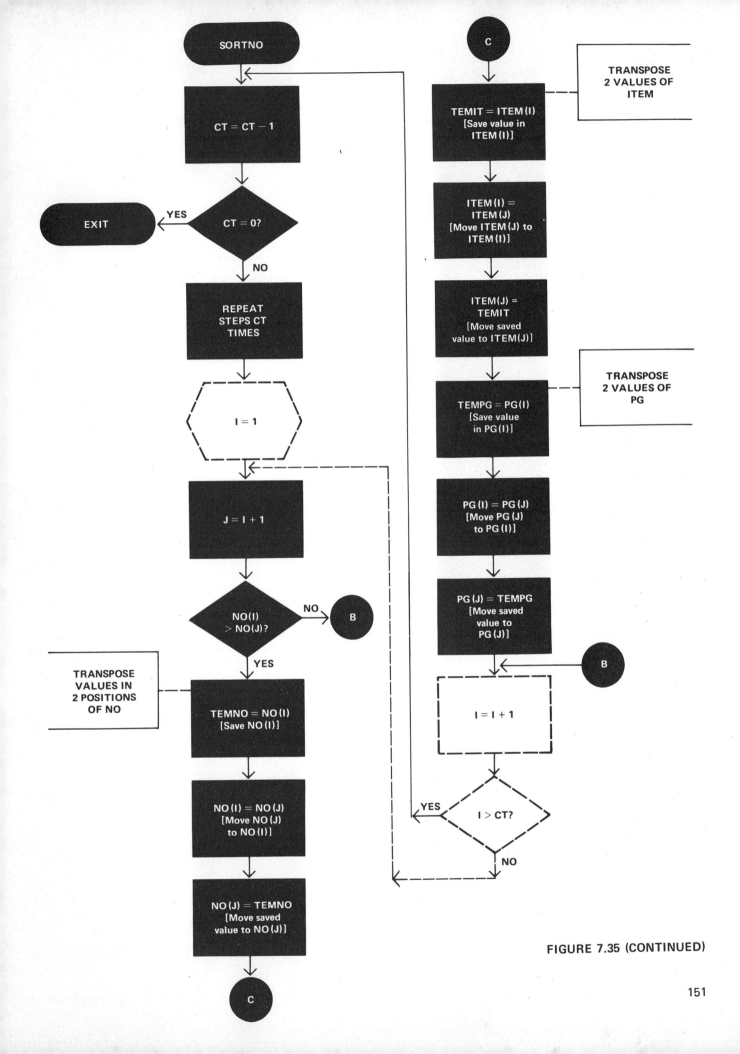

FIGURE 7.35 (CONTINUED)

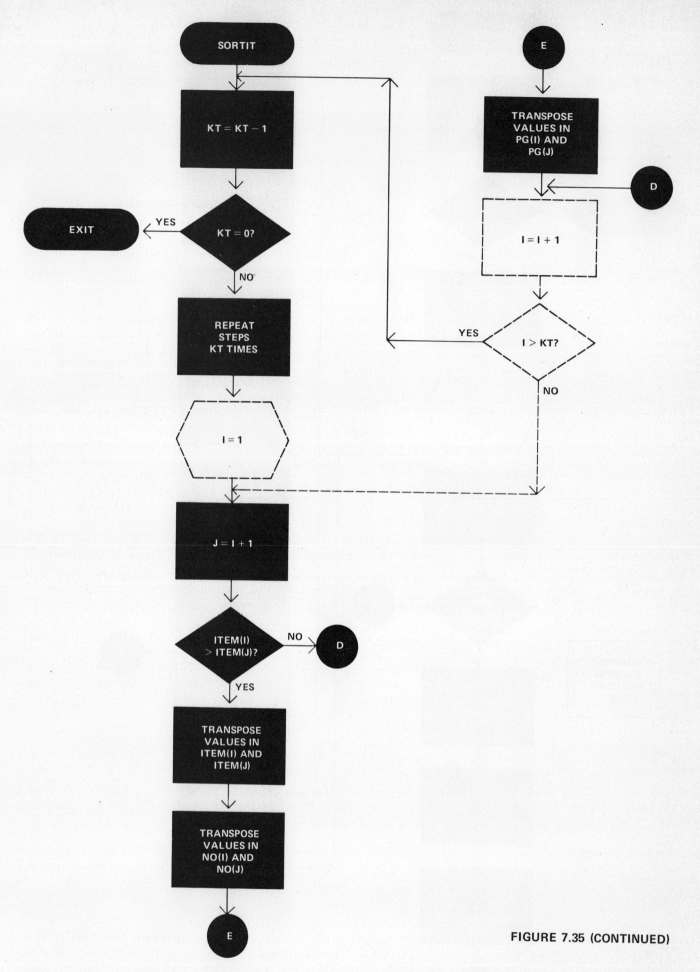

FIGURE 7.35 (CONTINUED)

Next, the program performs the WRITE routine, which prints out the order number index. The titles and column headings are printed out. Then a limited loop, executed C times (once for each record) prints out the sorted index entries. A counter, LINE, is included in the loop to limit the number of lines on a page to 25.

Then the program performs the second sort procedure (SORTIT) and rearranges the data alphabetically by item name. The last routine, WRITE, prints out the alphabetized index. The same algorithm is used for this index as the order number index, except that the item name is listed first on the page and then the order and page numbers are listed.

The sort procedure is based on a process of comparing two adjacent locations in an array and moving the larger value to the lower position. The sort routine SORTNO is composed of two nested loops. The inner loop is a limited loop and performs the comparisons of adjacent locations and transpositions of data. The outer loop, limited by a counter (CT), controls the number of times the sort process is repeated and limits the number of positions that are tested.

The routine begins by entering the outer loop. CT is decreased by one and tested to see if it is zero. If not, the program enters the limited inner loop. I, initialized to one, is the index for the first location to be compared, and J, initialized to two, is the index for the second. The data in the two positions are compared to see if the value on top is larger. If it is, they are transposed by moving the value in position one to a temporary location (TEMNO). Then the smaller value, in the second position, is moved to position 1. Then the larger value is moved from the temporary location to position 2. Now the related data in the two other arrays are transposed in the same way.

The index I is automatically increased by one and tested to see if it is greater than the current value of CT. If not, the program performs the inner loop again and compares the next two adjacent positions. If the value in position 1 is smaller than the value in position 2, no transposition takes place. The program automatically increases the index I, tests it, and returns to the beginning of the inner loop to compare the next two positions.

This process continues until the limit of the inner loop is reached. At this point all positions will have been tested. The largest value in the array will be in the bottom position.

Control returns to the beginning of the outer loop. CT is decreased by one since it is unnecessary to test the bottom position again. Control now reenters the inner loop and compares adjacent positions, transposing where necessary. When the inner loop terminates again, the second largest value in the array will be in the next to the last location. Control returns to the outer loop, CT is decreased, and the inner loop is repeated. This process continues until CT is equal to zero. At this point, the array will be in ascending order, the largest value at the bottom and the smallest at the top. (Descending sorts follow a similar algorithm.)

The alphabetic sort routine, SORTIT, follows a similar algorithm. A nested loop performs the comparisons and transpositions on the array ITEM. The related data in the NO and PG arrays are also transposed.

BUILDING BLOCKS USED:

No. 2 — Unconditional branch
No. 3 — Two-way conditional branch
No. 7 — Counters
No. 8 — Sequential loops
No. 9 — Nested loops
No. 10 — Terminating a loop with a trailer record
No. 12 — Terminating a loop with a counter
No. 13 — Limited loops
No. 14 — One-dimensional arrays

Exercises

1. Compare the placement of the counters in the READ and WRITE loops. Why is it necessary to subtract one from CT but not from LINE?
2. The sort routine SORTNO is composed of nested loops. What technique is used to control each loop?
3. What would happen if the limit of the inner limited loop were not decreased each time the outer loop executed? Would this be an error?
4. Redraw the flowchart of SORTNO to show two nested limited loops.
5. Redraw these same two nested loops using a positive counter in the outer loop and a negative counter in the inner loop.
6. Flowchart a program which would sort a group of values into descending numeric order.

UNIT 13   PERFORMING A BINARY SEARCH

PROBLEM   ● Galileo Apparatus Company sells a line of laboratory equipment, chemicals, and supplies to schools and colleges. They are in the process of designing a new system to facilitate data retrieval and maintenance. An inventory file, describing the 5000 items stocked by Galileo, is maintained by the company on punched cards. The data related to each item is recorded on a separate record:

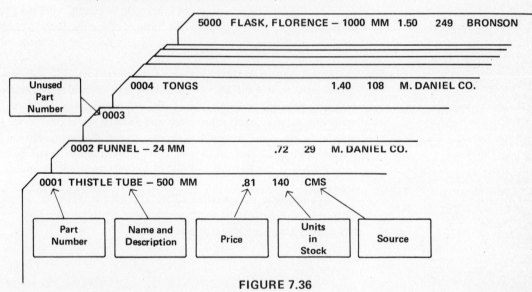

FIGURE 7.36

Personnel in the sales and order departments frequently refer to this file to check prices, availability of supply of various items, or other details. Activities in the inventory department and stockroom constantly generate information that must be incorporated into the records in the file.

As part of the new system, the data from the punched-card file will be transferred onto magnetic disk storage media to form an online data base accessible to the remote terminals located throughout the firm. This will allow Galileo to access the file quickly from several locations in the company.

As part of the new system, Galileo needs a program that will perform a search of the data base and allow the user to view and/or modify selected records. Output from the program will be on video screen and remote terminals.

SOLUTION • This problem illustrates an interactive program that uses a binary search to locate a specific record in a data file containing 5000 records. In a binary search, a file of records is continually divided in half until the desired record is located. The record is then either displayed on a video screen, or displayed and replaced by information input from the remote terminal. The program is structured to offer the user the option of repeating all or part of the program and is also designed to accommodate the addition of other options at a later time. (These might include preparing purchase orders for items that drop below a minimum level of supply, printing out suggestions for alternate or additional items to bring to the customer's attention, updating the file from the sales desk, or printing out notices to the stockroom to prepare the necessary quantity of items for shipment.)

The records in the data file are in sequential order by part number, and the file is recorded on a random access medium. Both are necessary prerequisites for a binary search. A binary search is preferable since it is faster than a sequential search, which checks each record in the file in sequence.

To begin, the program prints out the code numbers for the available options on the terminal and reads in the user's selection. Next it requests the user to enter the part number that is the object of search. The record that contains the relevant information is located in the data file and then the program branches to the procedure that performs the requested option.

If the user entered Code 1, the program branches to Point E and displays the record on the user's video screen. Then it queries the user on whether to display another record. An answer of YES branches the program back to read in another object of search. An answer of NO brings the user to the next option. This allows the user to repeat the entire program or to terminate execution.

A selection of Code 2 indicates the user wants to change all or part of a record. The program branches to Point F, and the record in question is displayed on the video screen. The user is asked to enter the code for the field to be changed and the new data. The program replaces all or part of the record with the updated information. The changed record is displayed on the video screen for confirmation and then the user is offered the option of repeating the sequence on another record. Finally, the user is offered the option of repeating the program or terminating execution. If an invalid code is entered the program repeats the query.

The routine SEARCH shows the algorithm followed in performing the binary search. The data file is referenced as though it were an array. (In fact, in many programs which perform binary searches, the data would first be read and stored in primary storage as an array.)

A binary search is based on the principle of continually dividing a file, or a portion of a file, in two and testing the record at the midpoint to see if it is the object of the search. If not, the program calculates whether the object is in the top or bottom portion, divides this in two, and tests the record at the midpoint. This process continues until the object is located.

For example, to locate the eighth record in a file of 12 records, the steps in a binary search would look like the figure at left. Only four records are tested. Eight would have to be tested in a sequential search.

The routine begins by initializing BOT to 1 and TOP to 5000. These define the size and numeric limits of the data file. Next, the program performs two error prevention tests. The object is compared against the part number stored in the first location of the file [PARTNO(BOT)]. If they are equal, the OBJ has been found and the record is moved to the work area accessible to the user. If the OBJ is a part number that is less than the one in this location, it indicates that the user made an error, and the program branches back to the input request.

Next the program tests to see if the OBJ is the part number at the top position of the file [PARTNO(TOP)]. If so, the search is ended and the record is made available to the user. If OBJ is higher than this part number, an input error has been made and the program again branches back to the input request.

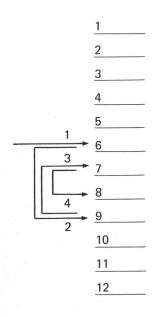

FIGURE 7.37

Chapter Seven: Applied Programming Logic

155

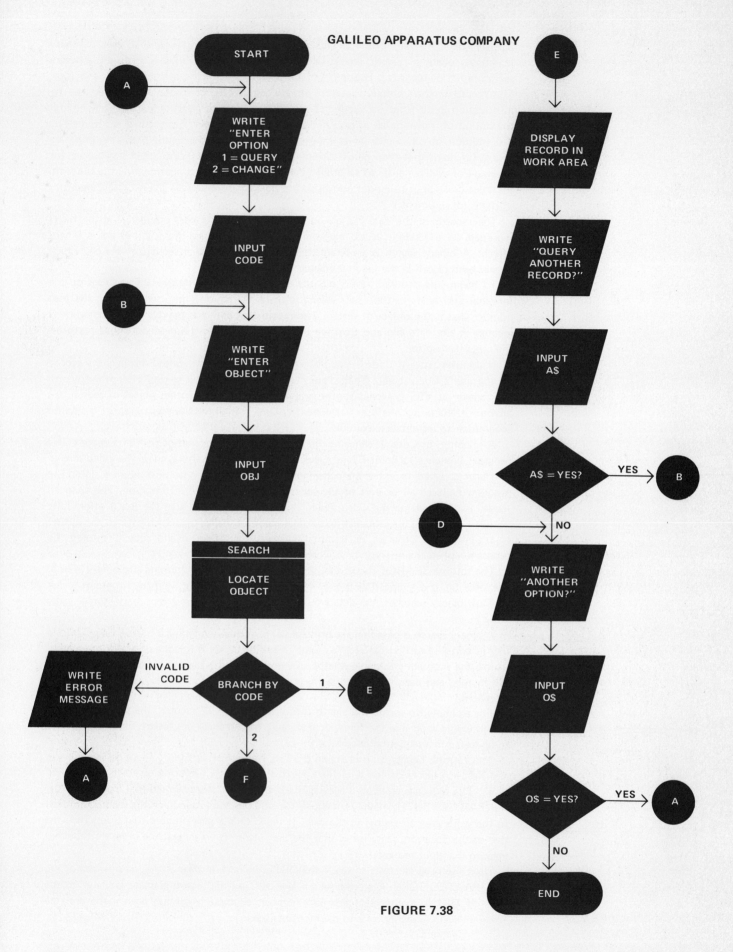

FIGURE 7.38

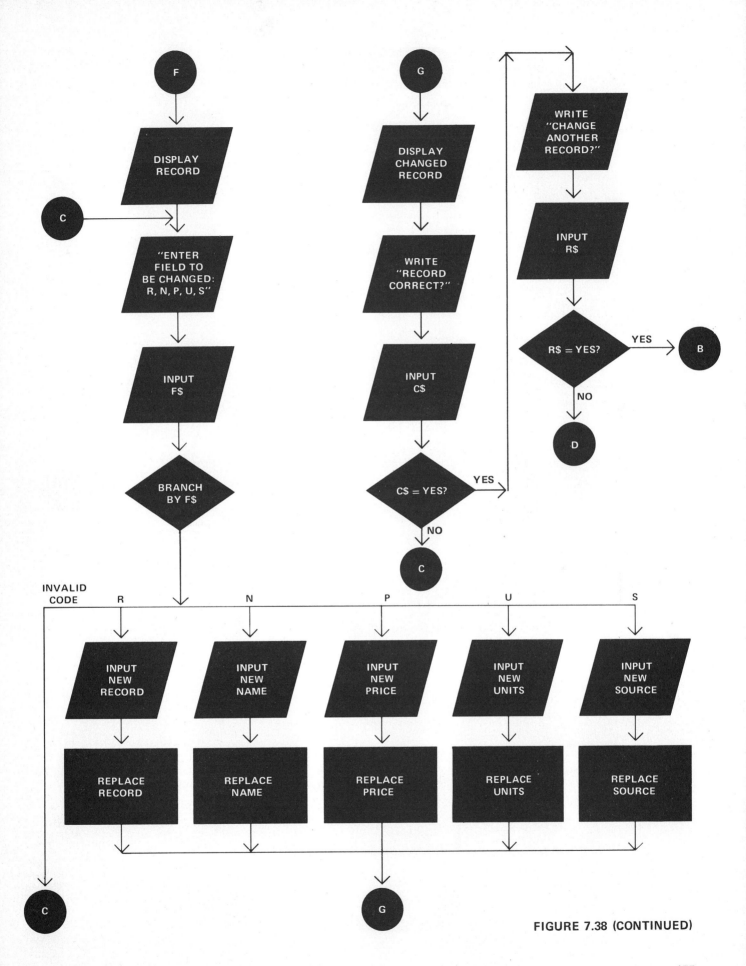

FIGURE 7.38 (CONTINUED)

Chapter Seven: Applied Programming Logic

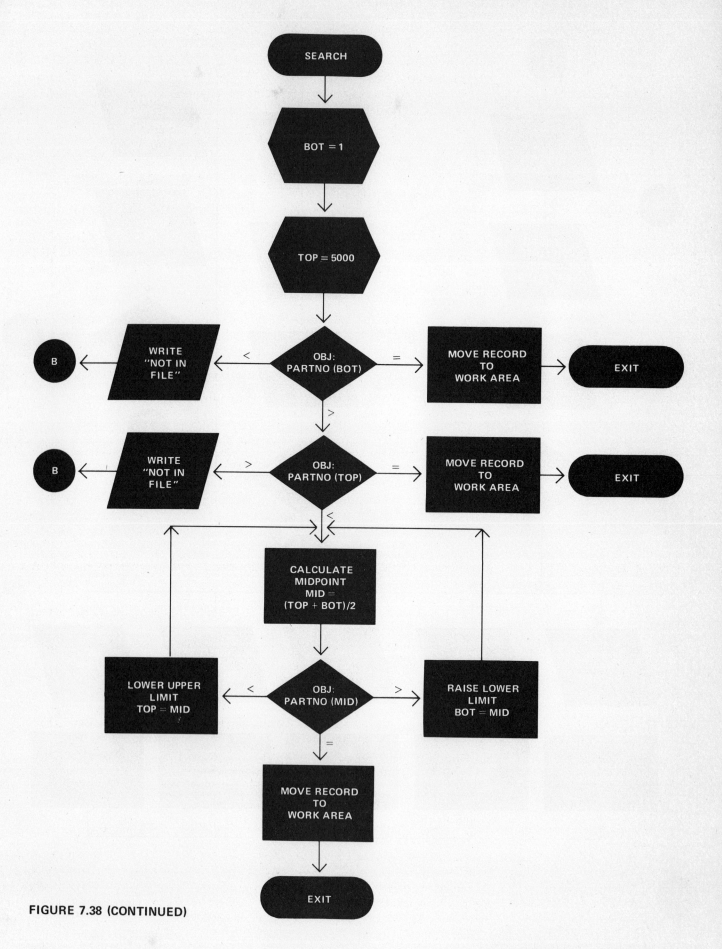

FIGURE 7.38 (CONTINUED)

Now the program begins the search proper. The midpoint of the file, MID, is calculated by adding TOP and BOT and dividing the sum by two. The part number stored in this location, PARTNO(MID), is compared to the OBJ. If they are equal, the search terminates and the record is made available to the user. If OBJ is less than PARTNO(MID), the OBJ is in the lower portion of the file and the top limit of the area being searched is lowered to the midpoint. If OBJ is greater than PARTNO(MID), the OBJ is in the upper portion of the file and the lower limit of the area being searched is raised to the midpoint.

The midpoint of this new smaller portion is now calculated by adding PARTNO(MID) to TOP or BOT and dividing the sum by two. The part number at this location is tested. This process continues until the OBJ is finally located and the search terminates.

Here are a few pointers to consider when programming binary searches. The formula used here to calculate the midpoint rounds down and the value in the top position will never be tested unless the program is specifically directed to test this one location. The mathematical step that calculates MID often produces a fractional number. Since MID serves as a subscript, only the integer portion of the number is used. In some languages, extracting the integer portion of a number requires a specific programming instruction.

In many files there are gaps between some of the numbers in the sequence. For example, the numbers in a file might be: 1, 2, 4, 5, 6, 8, 11, 12, etc. A program searching for a non-existent number could go into an endless loop. This can be avoided by counting the number of times the midpoint is calculated during each search, and limiting the repetitions to one-half the number of records in the file. If this limit is surpassed, an error message is printed out and the program branches back to the input request. In the file in this example, all part numbers between 1 and 5000 were included, whether or not they were assigned to an item. Therefore no test or limitations were required.

BUILDING BLOCKS USED:
No. 2 — Unconditional branch
No. 3 — Two-way conditional branch
No. 4 — Three-way conditional branch
No. 5 — Multiway conditional branch
No. 6 — Simple loop
No. 9 — Nested loops

Exercises

1. Can a binary search be used on all data input files? Explain.
2. In the search routine, would it be an error condition not to separately test for the part number in the bottom position?
3. Modify the SEARCH procedure to include an error routine to handle situations where the program is searching for a value not in the table.
4. Modify the program to offer the user the option of printing out selected data from the records.
5. Flowchart a program that reads in a card file with an undetermined number of records and performs a sequential search on the data. Assign a code number to each record to be used as the search field.
6. Using the file in Exercise 5, flowchart a program that sorts the data into numeric order and performs a binary search on the code field.

UNIT 14   FILE MAINTENANCE ROUTINE

PROBLEM ● Art Deco Magazine is mailed to thousands of subscribers monthly. Its mailing list must be updated, or modified, each month to include new subscribers, reduce the number of months until renewal by one, delete customers with expired subscriptions, and process renewals and address changes. Art Deco wants a computer program that will maintain its mailing list and prepare labels for the current month.

SOLUTION • File maintenance routines are frequently used in business to maintain groups of records to assure that they contain current, up-to-date information. File maintenance routines are used to update subscription lists, mailing lists, accounts receivable files, and inventory files. The essential characteristic of a file maintenance routine is that it merges one or more transaction files with a master file and produces a new, updated master file. Often, of course, it performs other procedures on the data as well.

A file maintenance routine was written for the magazine. It reads in a transaction file and a master file from two different devices and performs a sequence check on the records in the transaction file. It match/merges the records in the two files, writes the mailing labels, and then prepares a new master file.

Basically, this program consists of a series of conditional branches enclosed within a loop that has several entrance points. The branches direct the program to the appropriate processing procedures for each record and then back to the proper entrance point at the beginning of the loop to read in the next record.

Two input files are involved: the transaction file is on punched cards and the master file is on magnetic tape. The records in both files are processed one by one. The output file is on a second magnetic tape device.

Each record in the master file contains the name of the subscriber, the address, and the number of months remaining on the subscription (MTHS in the program):

NAME	ADDRESS		MONTHS
.....	.....	.....	...
.....	.....	.....	...
JENKINS, C.	8312 ALISON ST.	CANTON, OH 44720	5
JENSEN, L.	535 PLUMB AVE.	FALL RIVER, MA 02721	21
JOHNS, A.	5921 MANN DR.	EAU CLAIRE, WI 54701	0
JUAREZ, D.	32 W. M ST.	BURBANK, CA 91505	2
KANE, T.	666 SOUTH HILLS DR.	OMAHA, NE 68147	14
KIM, C.M.	3454 1ST ST.	COMPTON, CA 90220	1
KING, A.	1104 SPACE ST.	HOUSTON, TX 77007	8
.....	.....	.....	...
.....	.....	.....	...

FIGURE 7.39

This transaction file is alphabetized and contains the code, the name of the subscriber, the address, and the length of renewal period (R):

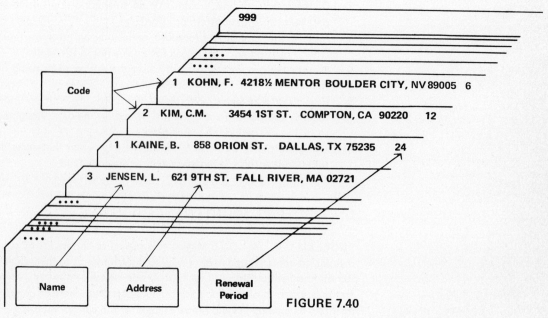

FIGURE 7.40

The code number on the transaction record tells the program the status of each record.
1 = New subscriber
2 = Subscription renewal
3 = Address change

The updated master file, recorded on a second magnetic tape device, will have this information:

. . . . .	. . . . .	. . . . .	. . .
. . . . .	. . . . .	. . . . .	. . .
JENKINS, C.	8312 ALISON ST.	CANTON, OH 44720	4
JENSEN, L.	621 9TH ST.	FALL RIVER, MA 02721	20
JUAREZ, D.	32 W. M ST.	BURBANK, CA 91505	1
KAINE, B.	858 ORION ST.	DALLAS, TX 75235	23
KANE, T.	666 SOUTH HILLS DR.	OMAHA, NE 68147	13
KIM, C.M.	3454 1ST ST.	COMPTON, CA 90220	12
KING, A.	1104 SPACE ST.	HOUSTON, TX 77007	7
KOHN, F.	4218½ MENTOR	BOULDER CITY, NV 89005	5
. . . . .	. . . . .	. . . . .	. . .
. . . . .	. . . . .	. . . . .	. . .

Mailing labels were prepared for all subscribers (in this portion of the file) except for Johns, A., whose name was deleted from the list. Two new subscribers were added, one address was changed, and one subscription was renewed.

The program begins by initializing TRR to A. This is used as a comparison field during the sequence checking procedure. Next TR, a record from the transaction file, is read in and tested to see if it is the trailer record. If it is not, the name on the record (TNAME) is compared to the value in TRR. If TNAME is less than TRR, the record is out of alphabetic order. The program branches to an error message and terminates. If TNAME is equal to or greater than TRR, TRR is assigned the current value of TNAME.

A record from the master file, MR, is read in and tested to see if it is the end of the file. If not, CD is assigned the value of the CODE field, read in from the last transaction record (point F in the flowchart). CD and CODE are used to branch the program to the appropriate routine for the record being processed.

Next are the match/merge procedures. TNAME and MNAME (from the master record) are compared to test their alphabetic relationship. If TNAME is greater, this indicates that the records do not match and should not be merged. It also indicates that MNAME comes first in the alphabet and will be processed first.

CD is reset to zero, which is the code assigned when no data is to be merged, and the program branches to point G to prepare the mailing label from the master record. At point G, the field for MTHS is tested to see if it is equal to zero. This would indicate an expired subscription and the program would branch to point D to read in the next master record. If MTHS is greater than zero, the program tests to see if it is less than four. If so, a subscription renewal notice is prepared. Next, the mailing label is written, using the information from the master record. MTHS is decreased by one, and a new master record is written on the output tape file.

The field that holds the master record (MR) is filled with blanks to indicate that no master record is being held for processing. The program tests CD to find the appropriate point at which to reenter the input cycle. Since a transaction record is still waiting in TR, the program reenters at point D to read in the next master record. Reentering at point A would have resulted in skipping the transaction record still in storage in TR.

The match/merge procedures are repeated on the new data. The new MNAME is compared to the TNAME still in storage. If they are equal, the names are the same. The program then tests the code and branches to the appropriate process for that set of records.

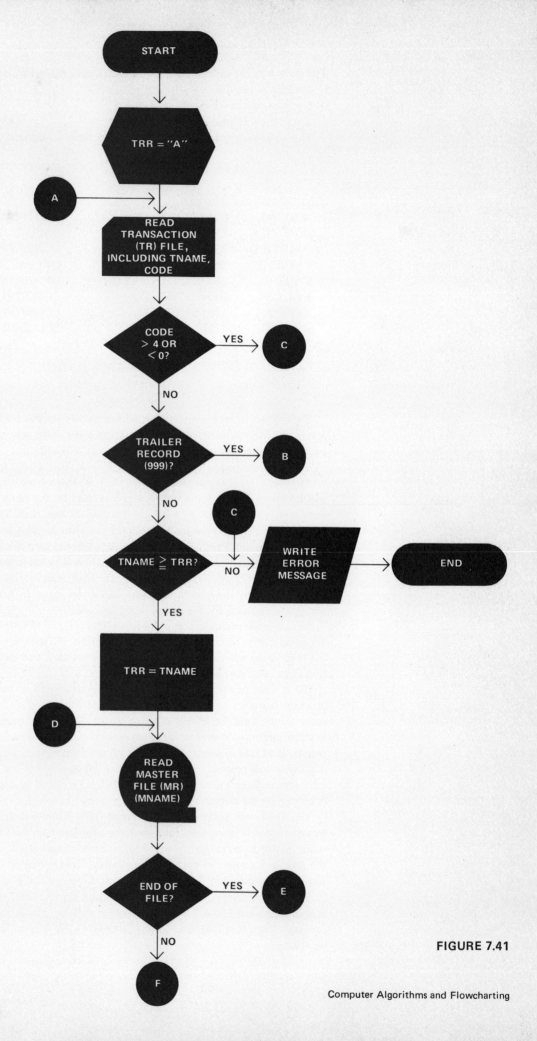

FIGURE 7.41

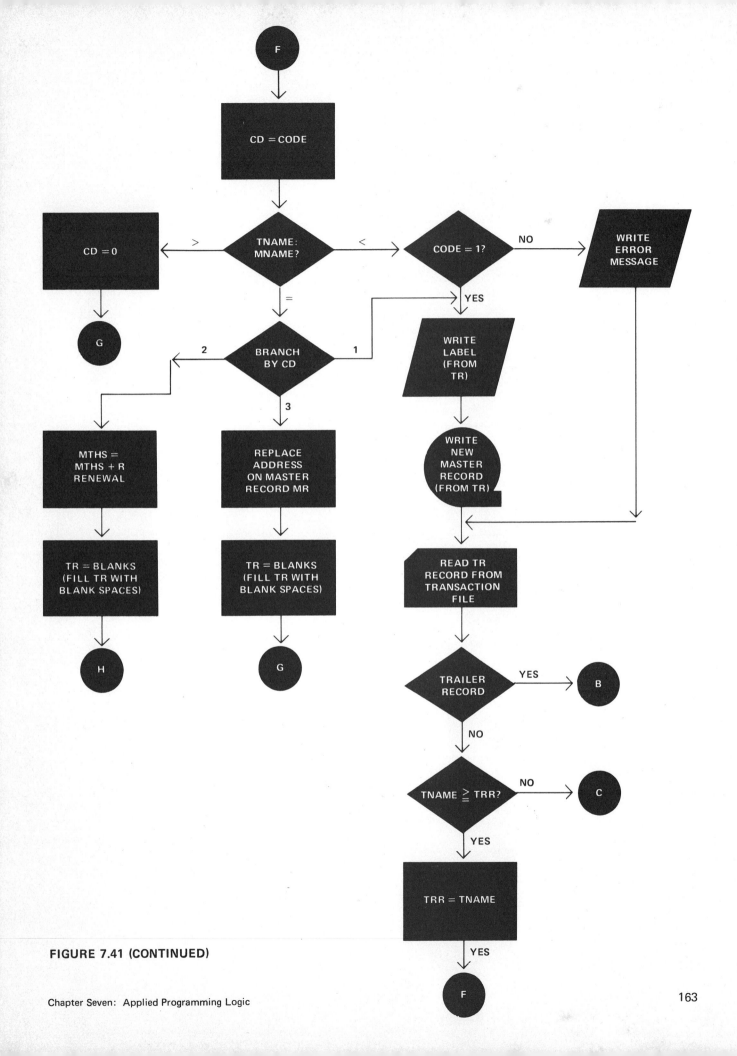

FIGURE 7.41 (CONTINUED)

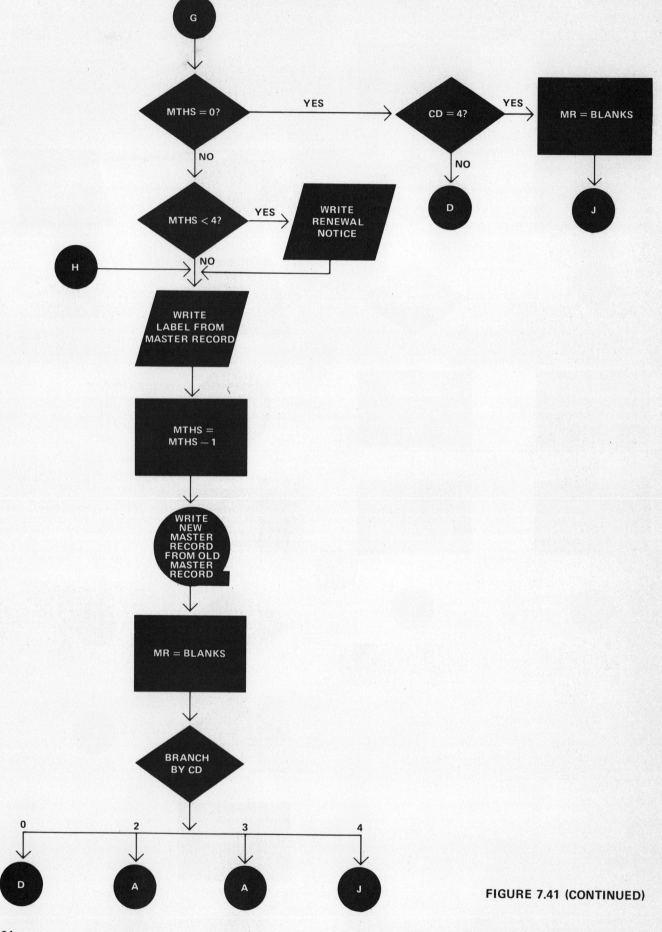

FIGURE 7.41 (CONTINUED)

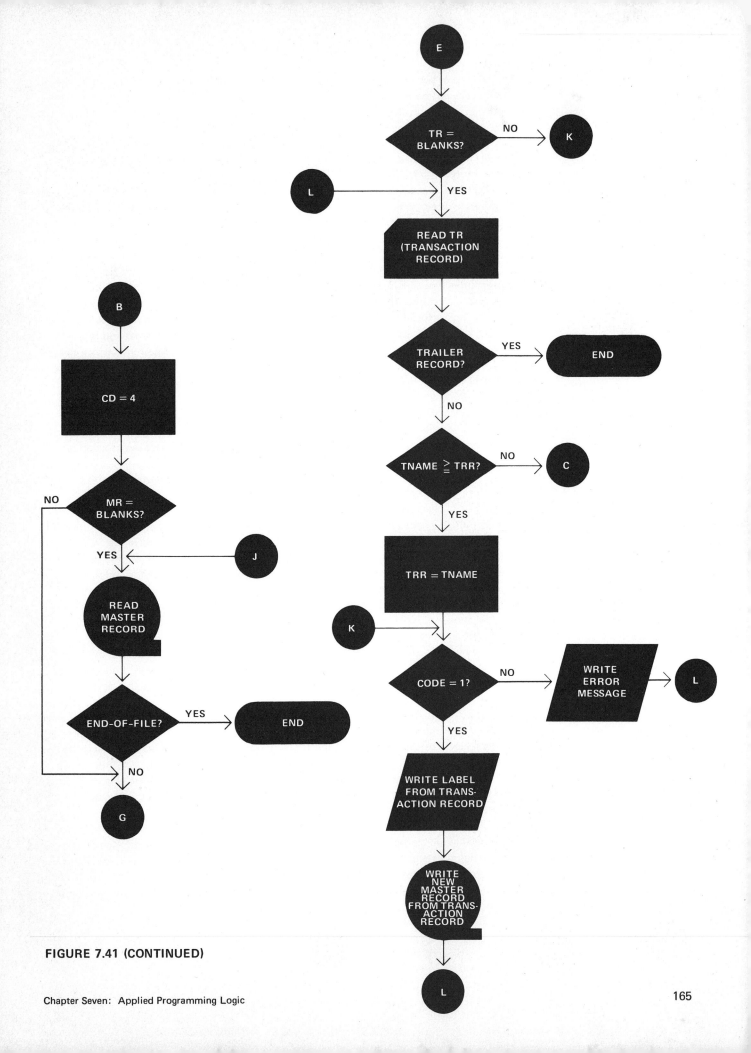

FIGURE 7.41 (CONTINUED)

Chapter Seven: Applied Programming Logic

A code of two indicates a renewal. The information on the transaction record is merged with that on the matching master record. TR is filled with blanks to indicate that there is no transaction record waiting in TR for processing. The program then moves to point H to write the mailing label and new master record. (Since this was a renewal, it was unnecessary to perform the tests at point G.) Since both master record and transaction record were merged and processed, the program reenters the input cycle at point A to read in two new records.

Matching names and a code of three read in from the transaction record indicate an address change. The information on the two records are merged and TR is filled with blanks. The program branches to point G to prepare the label and new master record. The program would again reenter the input cycle at point A since both records have been processed.

A code of one read in from the transaction record indicates a new subscriber. The program writes the mailing label and a master record from the information on the transaction record. Then the program reenters the input cycle at point F to read in a new transaction record only, since a master record is still in MR waiting to be processed.

If TNAME is less than MNAME, TNAME comes first in the alphabet and is processed first. The code is tested to see if it is one. If it is, it indicates a new subscriber and the appropriate procedure is carried out. If the code is not one, an error condition is present. The program writes an error message and then branches to read in the next transaction record.

The records in the two files are checked, matched, merged, and processed until the end of one of the files is reached. At this point, since there is only one file left to be processed, it is no longer necessary to perform the match/merge procedures and separate input cycle reentry points are used.

If the end of the transaction file is reached first, only the master file still contains unprocessed records. The program branches to point B. CD is set to four to direct the program to the correct input cycle reentry point. MR is checked to see if it holds blanks or a waiting record. A waiting record branches the program to point G and processing continues. If blanks are present in MR instead, a new master record is first read in and then processing continues at point G. The program reenters the input cycle at point J, which reads in only records from the master file. This continues until the end of the file is reached, and the program terminates execution.

If the end of the master file is reached first, only the transaction file contains records to be processed. The program branches to point E. At this point, all records left in the transaction file should be for new subscribers (CODE=1) or an error condition is present. The routine at point E first tests to see if there is a record waiting in TR. If there is, CODE is tested, the label and new master record are written from the information on the transaction record, and the program branches back to read in the next transaction record. When the end-of-file is reached, the program terminates execution.

Note, in the routine starting at point E, the program does not write blanks in TR after processing a record. The program must, therefore, return to point L rather than point E, or a logic error would occur — the program would enter an endless loop and the last records in the transaction file would not be processed.

A file maintenance program of this type that updates a mailing list of names would probably also include routines that perform additional error checks and could handle instances where several different subscribers have the same names.

Commands for rewinding tapes and opening and closing files would also have to be included in most programs. These will vary with the language being used for coding.

A file maintenance routine of this type could also be written to first match/merge the transaction and master files and write the updated master file. Then the tape with the new master file would be rewound back to the beginning, and the records would be read again to prepare the labels and notices.

BUILDING BLOCKS USED:

No. 2 — Unconditional branch
No. 3 — Two-way conditional branch
No. 4 — Three-way conditional branch
No. 5 — Multiway conditional branch
No. 6 — Simple loop
No. 10 — Terminating a loop with a trailer record

Exercises

1. Follow through the logic in the match/merge procedure, and explain why it was necessary to fill TR with blanks before branching to points H and G.
2. Why was it necessary to have a separate write and read sequence when the code was equal to one?
3. Modify the program to first match/merge the two files and write the updated master file, and then to prepare the labels and renewal notice.
4. Flowchart a program that merges three files that have matching records in sequential order.
5. Flowchart a program that match/merges two files based on account numbers.

Index

Addition, using a loop in, 67
Algorithms, 4, 12-15; *def.,* 4; *see also* Programs, building blocks for development of, 12-15
Alphabetic manipulation, 3, 91, 148-153; *illus.,* 148-152
　by arrays, 91
Alphanumeric characters, testing of, 63, 79
American National Standards Institute (ANSI), 9, 21, 29, 34, 91
Annotation symbol, 23, 32; *illus.,* 22
ANSI, *See* American National Standards Institute
APL, 5
Arithmetic and logic unit, 2
Arithmetic statements in COBOL, 70, 85
Arrays, 91-101, 119-120, 122, 128, 133, 153; *def.,* 91
　storage space needed for, 119-120
　one-dimensional, 91-96, 122, 128, 133, 153; *illus.,* 92-96
　　applications of, 122, 128, 133, 153
　two-dimensional, 97-101; *illus.,* 98-100
Arrowheads, 29
Assignment statement in FORTRAN, 70, 85, 107
Auxiliary operation symbol, 25; *illus.,* 22
Averages, calculation of, 67

BASIC, 5, 61-64, 66-67, 70, 74, 76, 80, 82, 85-86, 91-92, 97, 107, 147
　carriage control in, 147
　DIM statement in, 91, 92, 97
　FOR/NEXT statements in, 67, 74, 76, 86
　GO TO statement in, 61, 67, 74, 76,
　IF statement in, 62-64, 66, 80, 82, 85
　LET statement in, 70, 85
Batch programming, 137
Billings, programming monthly, 122-128; *illus.,* 125-127
Binary searches, 154-159; *illus.,* 154-158
Blank-record termination, *see* Sum-of-the-fields technique
Block diagram, 49; *see also* Modular programs, flowcharts for
Bonus incentive sales program, calculations for, 108-111; *illus.,* 108-110
Branches, 31-32, 60-61, 103, 107, 111-112, 118, 122, 128, 133, 144, 153, 159, 167
　conditional, *see* Conditional branches
　labeling of, 31; *illus.,* 32
　unconditional, *see* Unconditional branches

Business systems, 34-43
　designing of, 34-37
　expressed as system flowcharts, 37-41; *illus.,* 38, 40, 42-43

Carriage-control techniques, 147
Catalogues, indexing of, 148-53; *illus.,* 148-152
Categories, *see* Sorting
Central processing unit (CPU), 2, 5, 7
　in program execution, 5, 7
　storage in, 5
Checking account balances, processing of, 75; *illus.,* 77
Classification, *see* Sorting
COBOL, 5, 61-63, 66-67, 70, 74, 76, 80, 82, 85-86, 91, 97, 147
　arithmetic operation for adding in, 82
　arithmetic statements in, 70, 85
　carriage control in, 147
　GO TO statement in, 61, 67, 74, 76
　IF statement in, 62-63, 66, 80, 82, 85
　MOVE statement in, 70, 85
　OCCURS clause in, 91, 97
　PERFORM statement in, 67, 74, 76, 86
　VALUE clause in, 70, 85
Code numbers, testing, 65
Coding, 4-5, 62, 64, 66, 122; *def.,* 4
　of conditional branches, 62, 64, 66
　of several sequential limited loops, 122
Coding sheet, 5; *illus.,* 6
Coefficient of correlation, 133-137; *def.,* 133
　program to determine, 133-137; *illus.,* 134-135
Collate symbol, 26; *illus.,* 22
Comment symbol, *see* Annotation symbol
Communication links, 25, 35 *illus.,* 23
　in business systems, 35
　symbol, 25
Compilation, *def.,* 5
Compiler (translator program), 5
Computer logic, 3, 14
　symbols for, 14
Computer system, 2-3; *illus.,* 2
Computer time and sequential loops, 74
Conditional branches, 61-66, 80, 82-83, 85, 103, 107, 111, 112, 122, 128, 133, 144, 153, 159, 167
　coding of, 62, 64, 66
　multiway, 65-66, 82-83, 85, 112, 159, 167; *illus.,* 66
　　applications of, 112, 159, 167
　　limitation of executions in, 82, 85; *illus.,* 83
　　to indicate terminating a loop with a trailer, 80
　three-way, 63-64, 111-112, 118, 159, 167; *illus.,* 64
　　applications of, 111-112, 118, 159, 167
　two-way, 61-62, 103, 107, 111-112, 118, 122, 128, 133, 144, 153, 159, 167; *illus.,* 62
　　applications of, 103, 107, 111-112, 118, 122, 128, 133, 144, 153, 159, 167

Connector symbols, 26, 29, 31, 67, 74, 76; *illus.,* 22
　to indicate looping, 67, 74, 76
Consecutive numbering, 69
Consumer credit system, flowchart of, 39-41; *illus.,* 40
CONTINUE statement, 86
Control unit, 2
Core symbol, 24; *illus.,* 23
Correlation analysis, 133-137; *def.,* 133
　programmed, 133-137; *illus.,* 134-135
Cost analyses, 36
Costs, calculations from lists of, 72
Counters, 69-72, 82-85, 92-93, 97, 118, 128, 133, 144, 153; *illus.,* 70, 71
　applications of, 118, 128, 133, 144, 153
　incorrect placement of, *illus.,* 71
　to load and manipulate arrays, 92, 97; *illus.,* 93
　to terminate loops, 82-85, 92, 153; *illus.,* 83-84, 93
　　application of, 153
CPU, *see* Central processing unit
Cross-reference symbols, 33; *illus.,* 33

Data base, *def.,* 35
Data file, *see* Files
Debugging, *def.,* 7
Decision points, labeling of, 31; *illus.,* 32
Decision symbol, 25; *illus.,* 22
Decision tables, 11-12, 122-128; *def.,* 4; *illus.,* 11, 13, 123
Descriptive text within symbols, 32; *illus.,* 32; *see also* Annotation symbol
Different sets of data, using unconditional branches to process, 60
DIM statement, 91-92, 97
DIMENSION statement, 91, 97
Direct observation, 36
Discounts, calculating, 97
Display symbol, 24; *illus.,* 23
DO statement, 67, 74, 76, 86
Document symbol, 24; *illus.,* 23
Documentation, *def.,* 7

END, 26, 52, 59
Error messages, branching to, 61
Error routine, sum-of-the-fields technique as resulting in, 80
Execution, 5, 7; *def.,* 5
EXIT, 26, 52
Extract symbol, 26; *illus.,* 22

Fact-finding techniques, 36
Files, 72-73, 105-111, 128-133, 159-167; *see also* Arrays; Items
　processing, 105-111, 128-133
　　by groups of records, 108-111; *illus.,* 108-111
　　multiple input/output, 128-133; *illus.,* 129, 131-132
　　one record at a time, 105-107; *illus.,* 105-106

Computer Algorithms and Flowcharting

Files *(continued)*
    routine for maintaining, 159-167; *illus.,* 160, 162-165
    sequential loops in manipulating, 72-74; *illus.,* 73
Flight reservation system, online, flowchart of, 41; *illus.,* 43
Flowcharts, 4, 15-33, 47-57; *def.,* 4; *illus.,* 4-5, 16; *see also* Programs, building blocks for
    designing, 29-33
    detail program, 4, 18-19, 49, 57; *def.,* 4; *illus.,* 19, 48, 50
    modular program, 4, 18, 49, 57; *def.,* 4; *illus.,* 18, 47, 49, 54-55
    preparing, 21-33, 49-57
    system, *see* System flowcharts
Flowlines, 21, 23, 29, 67, 74, 76; *illus.,* 22
    to indicate looping, 67, 74, 76
FOR/NEXT statements, 67, 74, 76, 86
Formulas, 32, 133-137
    programming of, 133-137; *illus.,* 134-135
    in symbols, 32; *illus.,* 32
FORTRAN, 5, 61-63, 66-67, 70, 74, 76, 80, 82, 85-86, 91, 97, 107, 147
    arithmetic operation for adding in, 82
    assignment statement in, 70, 85, 107
    carriage control in, 147
    CONTINUE statement in, 86
    DIMENSION statement in, 91, 97
    DO statement in, 67, 74, 76, 86
    GO TO statement in, 61, 67, 74, 76
    IF statement in, 62-63, 66, 80, 82, 85
    INT function in, 107

GO TO statement, 61, 67, 74, 76
Graphics, output, 51-52; *illus.,* 51
Graphs, preparation of, 144-147; *illus.,* 145-146

IBM flowcharting worksheet, 29; *illus.,* 30
IBM template, 27-29; *illus.,* 28
IF statement, 62-63, 66, 80, 82, 85
IF/THEN relationship, 12
Indexes (subscripts), 91, 97; *def.,* 91
Input, 2, 14, 36
    in system design, 36
Input/output files, processing multiple, 128-133; *illus.,* 129, 131-132
Input/output symbols, 21, 23-25; *illus.,* 22-23
    specialized, 23-25
Instructions, sequences of, *see* Sequences
Interactive programming techniques, 137-144, 154-159; *illus.,* 138-143
    using binary search, 154-159; *illus.,* 154, 156-158
Interest, calculation of, 59, 137-144; *illus.,* 138-143
International Organization for Standardization (ISO), 21
INTERRUPT, 26
Interviews, 36

Inventory file, binary search of, 154-159; *illus.,* 154, 156-158
ISO (International Organization for Standardization), 21
Items, 61, 75, 78; *see also* Files
    loop for undetermined number of, 78
    repetitive operations on, 75
    selecting specific, 61

Keyboarding, 5
Keypunching, 5

Labeling in flowcharts, 31; *illus.,* 32
Languages, 4, 5; *see also* APL; BASIC; COBOL; FORTRAN; PL/1
    machine, *def.,* 4
    problem-oriented (POLs), 5
LET statement, 70, 85
Letters, loop preparation of, 67
Limited loops, *see* Loops, limited
Lines, numbering of, 82, 85; *illus.,* 83
Location numbers, 97
Logic, *see* Arithmetic and logic unit; Computer logic
Loop termination, 67, 78-85, 92-93, 103, 107, 111-112, 118, 128-133, 153, 167
    sum-of-the-fields technique in, 80-82; *illus.,* 81
        application of, 111
    trailer record(s) in, 78-80, 82, 84-85, 103, 107, 111-112, 118, 128, 133, 153, 167; *illus.,* 78-80
        applications of, 103, 107, 111-112, 118, 128, 133, 153, 167
        counters combined with, 82, 85; *illus.,* 84
    counters in, 82-85, 92; *illus.,* 83-84, 93
        application of, 153
Loops, 67-68, 72-77, 86-95, 97, 103, 107, 111-112, 118-122, 128, 133, 137, 144, 147, 153, 159, 167; *see also* Loop termination
    limited, 86-95, 97, 119-122, 128, 133, 137, 144, 147, 153; *illus.,* 87-89, 92, 94-95
        applications of, 122, 128, 133, 137, 144, 147, 153
        for arrays, 91-95, 97, 119-122; *illus.,* 92, 94-95, 119-121
    nested, 74-77, 89-90, 97, 111, 118, 147, 153, 159; *illus.,* 76-77
        applications of, 111, 118, 147, 153, 159
        limited, 90, 97; *illus.,* 89
    sequential, 72-74, 88, 90, 119-122, 128, 133, 147, 153; *illus.,* 73
        applications of, 122, 128, 133, 147, 153
        limited, 90, 119-122; *illus.,* 88, 119-121
    simple, 67-68, 103, 107, 112, 144, 159, 167; *illus.,* 68
        applications of, 103, 107, 112, 144, 159, 167

Machine language; *def.,* 4; *see also* Languages
Magnetic disk symbol, 24; *illus.,* 23
Magnetic drum symbol, 24; *illus.,* 23
Magnetic tape symbol, 24; *illus.,* 23
Mailing list, maintenance of, 159-167; *illus.,* 160-165
Manual input sumbol, 24; *illus.,* 23
Manual operation symbol, 25; *illus.,* 22
Mathematical operations, 3, 14, 59
Matrix manipulations, two-dimensional arrays for, 97
Merge symbol, 25; *illus.,* 22
Modular programs, 4, 18, 45-47, 49, 54-55, 57; *def.,* 45
    flowcharts for, 4, 18, 49, 57; *def.,* 4; *illus.,* 18, 47, 49, 54-55
MOVE statement, 70, 85
Multiway conditional branch, 65-66, 82-83, 85, 112, 159, 167; *illus.,* 66
    applications of, 112, 159, 167
    limitation of number of executions in, 82, 85; *illus.,* 83

Nested loops, 74-77, 89-90, 97, 111, 118, 147, 153, 159; *illus.,* 76-77
    applications of, 111, 118, 147, 153, 159
    limited, 90, 97; *illus.,* 89
Numeric sorts, 148-153; *illus.,* 148-152

Observation, direct, 36
OCCURS clause, 91, 97
Offline storage symbol, 25; *illus.,* 23
One-dimensional arrays, 91-96, 122, 128, 133, 153; *illus.,* 92-96
    applications of, 122, 128, 133, 153
Online, *def.,* 137
Online flight reservation system, flowchart of, 41; *illus.,* 43
Online storage symbol, 24; *illus.,* 23
Output, 2-3, 14, 36, 51-52
    designing graphic, 51-52; *illus.,* 51
    in system design, 36
Output forms, numbering of, 69

Page continuators, 31
Pages, numbering of, 69, 82, 85; *illus.,* 83
Parallel mode, *illus.,* 22
Parts in stock, 63, 111-114
    classifying defective, 111-114; *illus.,* 111-114
    testing the number of, 63
Payments, sorting of, 63
Payroll, calculation of, 128-133; *illus.,* 129, 131-132
PERFORM statement, 67, 74, 76, 86
Personnel considerations in system design, 36
Personnel report, preparation of, 115-118; *illus.,* 115-117
PL/1, 5
POLs (problem-oriented languages), 5
Predefined process symbol, 25; *illus.,* 22
Preparation symbol, 25; *illus.,* 22

Index   169

Primary storage unit, 2, 5
    instruction storage in, 5
PRINT statement, 90
Print-outs, 65, 115-118
    of messages according to category, 65
    with literal text, 115-118; *illus.,* 115-117
Problem analysis, 3-4, 9-12; *def.,* 9
Problem-oriented languages (POLs), 5
Problems, defining, 34-37
Process symbols, 21, 25-27; *illus.,* 22
Programmers, skills required of, 7-8
Programs, 1, 3-7, 59-101; *def.,* 1
    building blocks for, 59-101; *see also* Arrays; Conditional branches; Loop termination; Loops; Single-pass execution; Unconditional branches
    detail, flowcharts for, *see* Flowcharts, detail program
    examples of, 91
    modular, flowcharts for, *see* Flowcharts, modular program
    planning and developing, 3-7
    repeating of entire, 60
    translator (compilers), 5
Pulses, electronic, 2
Punched-card symbol, 23-24; *illus.,* 23
Punched-tape symbol, 24; *illus.,* 23

Questionnaires, 36

Rapidesign, Inc., template, *illus.,* 27
Real estate listing system, flowcharts for, 41, 57; *illus.,* 42, 58
Real time, *def.,* 137
Reference symbols, 33; *illus.,* 33
Relational comparisons and decisions, 61
Relational operators, 65
Relationships, establishment of, 11-12
Repeating a process, *see* Loops
Repetitions, setting a limit on, 82; *illus.,* 83
Reports, preparation of, 115-118; *illus.,* 115-117
Rule, *def.,* 12

Sales order system, flowcharts of, 39, 57; *illus.,* 38, 56
Searches, 59, 75, 97, 154-159
    binary, 154-159; *illus.,* 154-158
Secondary storage, 2

Sequences, 60-61, 67, 74
    directing control to alternate, 60
    repeating, 60, 61, 67, 74
    skipping, 60, 61
Sequential loops, 72-74, 88, 90, 119-122, 128, 133, 147, 153; *illus.,* 73
    applications of, 122, 128, 133, 147, 153
    limited, 90, 119-122; *illus.,* 88, 119-122
Simple loop, 67-68, 103, 107, 112, 144, 159, 167; *illus.,* 68
    applications of, 103, 107, 112, 144, 159, 167
Single-pass execution, 59, 137; *illus.,* 59
    application of, 137
Sort symbol, 26; *illus.,* 22
Sorting, 61, 63, 65, 75, 91, 97, 111-114, 148-153
    into many categories, 65
    by nested loops, 75
    numeric and alphabetic, 148-153; *illus.,* 148-152
    by one-dimensional arrays, 91
    into three categories, 63, 111-114; *illus.,* 111-114
    into two categories, 61
    by two-dimensional arrays, 97
Specifics, reducing problem to, 10-11
START, 26, 52, 59
Statistical analyses, 36
Stock, parts in, 63, 111-114
    classifying defective, 111-114; *illus.,* 111-114
    testing of number of, 63
STOP, 26
Storage, 2, 5, 59
    primary and secondary, 2
    printing a block of data from, 59
Strategies, 12; *see also* Algorithms
Striped symbols, 33; *illus.,* 33
Subroutines, 33, 46-49, 65
    by category, 65
    flowcharts for, *illus.,* 47
    indicated by striped symbols, 33; *illus.,* 33
Subscription file, maintaining, 159-167; *illus.,* 160-165
Subscripts (indexes), 91, 97; *def.,* 91
Sum-of-the-fields technique, 80-82, 111; *illus.,* 81
    application of, 111
Symbols, 14, 21-27, 29, 32-33
    in flowchart preparation, 21-27, 29, 32-33, 41; *illus.,* 22-23
    for logical operations, 14
    for mathematical operations, 14
System, 34, 36-7; *def.,* 34

System flowcharts, 4, 18, 34, 37-44; *def.,* 4; *illus.,* 17
    business systems expressed as, 37-41; *illus.,* 38, 40, 42-43
    preparing, 41-44
Systems analysis, 34-44
    goals of, 35
    preparing flowcharts in, 41-44
    scientific method in, 35-37
Systems analysts (systems engineers), *def.,* 34

Tables, 4, 11-13, 97, 122-128
    decision, 4, 11-12, 122-128; *def.,* 4; *illus.,* 11, 13, 123
    two-dimensional arrays in manipulation of, 97
Tax bills, computing, 59
Templates, 27-29; *illus.,* 27-28
Terminal symbol, 26; *illus.,* 22
Termination, *see* Loop termination
Tests by conditional branches, 61-66
Three-way conditional branch, 63-64, 111-112, 118, 159, 167; *illus.,* 64
    applications of, 111-112, 118, 159, 167
Trailer record, 78-80, 82, 84-85, 103, 107, 111-112, 118, 128, 133, 153, 167; *illus.,* 78-80
    applications of, 103, 107, 111-112, 118, 128, 133, 153, 167
    counters combined with, 82, 85; *illus.,* 84
Two-dimensional arrays, 97-101; *illus.,* 98-100
Two-way conditional branch, 61-62, 103, 107, 111-112, 118, 122, 128, 133, 144, 153, 159, 167; *illus.,* 62
    application of, 103, 107, 111-112, 118, 122, 128, 133, 144, 153, 159, 167

Unconditional branches, 60-61, 103, 107, 111-112, 118, 122, 128, 133, 144, 153, 159, 167; *illus.,* 60
    applications of, 103, 107, 111-112, 118, 122, 128, 133, 144, 153, 159, 167

VALUE clause, 70, 85
Variables, 10
    definition and quantification of, 10
    test, *see* Tests by conditional branches

Worksheets, flowcharting, 29; *illus.,* 30